
ADVANCED   NATIONAL   SEISMIC   SYSTEM

ShakeMap® Manual

TECHNICAL MANUAL, USERS GUIDE, AND SOFTWARE GUIDE

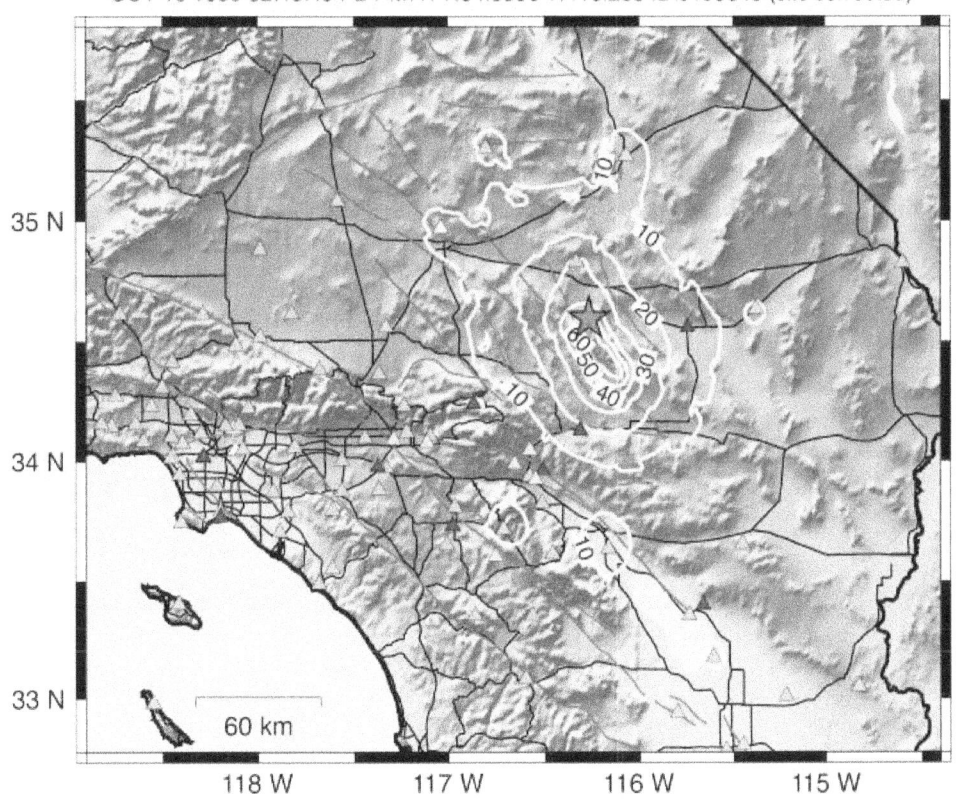

Trinet Peak Accel. Map (in %g) for Hector Mine Earthquake
OCT 16 1999 02:46:45 PDT M7.1 N34.5956 W116.268 ID:9108645 (site corrected)

Prepared by

David J. Wald, Bruce C. Worden, Vincent Quitoriano, and Kris L. Pankow

FOREWORD

ShakeMap (http://earthquake.usgs.gov/shakemap)—rapidly, automatically generated shaking and intensity maps—combines instrumental measurements of shaking with information about local geology and earthquake location and magnitude to estimate shaking variations throughout a geographic area. The results are rapidly available via the Web through a variety of map formats, including Geographic Information System (GIS) coverages. These maps have become a valuable tool for emergency response, public information, loss estimation, earthquake planning, and post-earthquake engineering and scientific analyses. With the adoption of ShakeMap as a standard tool for a wide array of users and uses came an impressive demand for up-to-date technical documentation and more general guidelines for users and software developers. This manual is meant to address this need.

ShakeMap, and associated Web and data products, are rapidly evolving as new advances in communications, earthquake science, and user needs drive improvements. As such, this documentation is organic in nature. We will make every effort to keep it current, but undoubtedly necessary changes in operational systems take precedence over producing and making documentation publishable. As this report is published through the USGS, the sole location of this manual is at Web Uniform Resources Locator (URL):

http://pubs.usgs.gov/tm/2005/12A01/

Some sections or subsections of the manual are seemingly incomplete. However, we have purposely included section or subsection headings as placeholders for products in development or regional ShakeMap information so that the user is aware of its existence and ongoing development. In these circumstances we simply mark the section with [**TBS**], for "to be specified."

Please address and any concerns or specific questions about this documentation to the ShakeMap Working Group via the ShakeMap Web page comment form.

TABLE OF CONTENTS

INTRODUCTION AND OVERVIEW

The most common information available immediately following damaging earthquakes has traditionally been their magnitude and epicentral location. However, the damage pattern is not a simple function of these two parameters alone, and more detailed information is necessary to properly evaluate the situation. ShakeMap® has proven to be a useful, descriptive display for rapidly assessing the scope and extent of shaking and potential damage following an earthquake.

ShakeMap's production of the maps is automatic, triggered by any significant earthquake in an area of the country where the ShakeMap system is in place. Maps are made available within several minutes of the earthquake for public and scientific consumption via the World Wide Web; they will be made available with dedicated communications for emergency response agencies and critical users. Such maps have traditionally been difficult to produce rapidly and reliably due to limitations of seismic network instrumentation and data telemetry. In addition, adequate relationships between recorded ground-motions and damage intensities have only recently been developed. However, with recent advances in digital communication and computation, it is now technically feasible to develop systems to display ground-motions in an informative manner almost instantly.

We generate separate maps of the spatial distribution of peak ground-motions (acceleration, velocity, and spectral response) as well as a map of instrumentally derived seismic intensities. These maps provide a rapid portrayal of the extent of potentially damaging shaking following an earthquake and can be used for emergency response, loss estimation, and for public information through the media. For example, maps of shaking intensity can be combined with databases of inventories of buildings and lifelines to rapidly produce maps of estimated damage. A detailed description of the shaking over a large region requires interpolation of measured ground-motions unless the recordings are extremely abundant. In the ShakeMap implementation, empirically based ground-motion estimation combined with simple geologically based, frequency and amplitude-dependent site correction factors provide a useful first-order correction for local amplification in areas that are not instrumented.

In this manual we describe the current ShakeMap system and implementation as well as ongoing operational and development efforts pertinent to ShakeMap under the Advanced National Seismic System (ANSS). ShakeMap was originally designed to be a Web-based information system; so much of its functionality and utility is fundamentally integrated into its Web pages. However, a number of other ShakeMap-related products are now available.   In Section 1, the *Users' Guide*, these products and their methods for delivery and use are fully outlined.   In Section 2, the *Technical Manual*, the production of the ShakeMap and its associated products is explained in detail, providing users the necessary background to understand the derivation of each product thereby assuring the most appropriate uses and decision making practices. Because the ShakeMap software has been ported to a number of regions within the United States as well as in other countries, we also include Section 3, a *Software Guide,* which provides an introduction to the ShakeMap software package, including background and guidance for installation and operation.

An overview of the contents of these manuals is provided below. There is some redundancy among these three sections, in particular between the *User's Guide* and the *Technical Manual*, because the intent and likelihood is that, as Web-based manuals, these will be downloaded and used independently.

In the ***Users' Guide***, we describe basic ShakeMap products and their current and potential uses. First, we provide an overview of current ShakeMap applications. We then explain the different formats and types of maps available and describe the ShakeMap Web pages. Next, we expand on different automated mechanisms to receive ShakeMap, including new approaches undergoing further development, particularly ShakeCast. We also describe Scenario Earthquake ShakeMaps, which provide the basis for pre-earthquake planning and understanding the potential effects of large earthquakes in the future. In each subsection, we try to provide concrete examples of potential uses of each product as well as notable users for each example. Although we show several ShakeMap Web page examples in the *User's Guide*, this guide is no substitute for the ShakeMap Web pages, and we recommend having a Web browser open to those pages while the *User's Guide* is in hand.

The ***Technical Manual*** is meant as the definitive source of information pertaining to the generation of ShakeMaps. The initial description of Wald and others (1999a) is outdated and is superseded by this manual. In the *Technical Manual*, we detail the approaches used for gap filling between stations by employing predictive ground-motion relationships, interpolation using inferred site amplifications, and the conversion of ground-motion recordings to instrumental intensity. We also provide background and some justifications for the choice of the ground-motion parameters mapped and describe both the data acquisition and processing procedures. The approach used for generating Earthquake Scenario ShakeMaps (used for response planning purposes) and Composite ShakeMaps (combining predictive ground-motions, observed ground-motions, and historic or other macroseismic intensities) is also detailed.

Finally, in order to enable customization for specific earthquakes or for different regions of the United States, each ShakeMap module has an accompanying collection of configurable parameters set in separate configuration files. For example, in these files one assigns the regional boundaries and mapping characteristics to be used by the Generic Mapping Tool (GMT), where and how to transfer the maps, email lists and file delivery lists, and so on. Specific details about the software and configuration files are described in detail in the *Software Guide*.

Technical users of ShakeMap should, however, also consult the *User's Guide* for additional information pertaining to the format, availability, and the range of ShakeMap related products that are available.

The ***Software Guide*** provides an overview of the ShakeMap software package for current and potential users of the software, and includes both the necessary background and guidance for ShakeMap installation and operation. ShakeMap is a collection of programs, largely written in the PERL programming language. These programs are run sequentially to produce ground-motion maps as well as Web pages and pager/email notifications. In addition to PERL, a number of other software packages are used. In keeping with our development philosophy, all additional software components are built from freely available, open-source packages.

PERL is a powerful, freely available scripting language that runs on all computer platforms. The collection of PERL modules allows the processing to flow in discrete steps that can be run collectively or individually. Within the PERL scripts, other software packages are called, specifically packages that enable the graphics. For instance, maps are made using the Generic Mapping Tool (GMT; Wessel and Smith, 1991). Parametric and earthquake-specific data and mapping parameters are stored and queried via MySQL databases, and much of the Web and parametric data handling is done with XML tagging.

With recent advances in GIS software and usage, several aspects of the ShakeMap system could be accomplished within GIS applications, but the open-source, freely available nature of GMT combined with PERL scripting tools allows for a flexible and readily available ShakeMap software package. Nonetheless, we do take advantage of GIS for a number of products as described in the *User's Guide*.

MESSAGE TO USERS

ShakeMap is designed to rapidly produce shaking and intensity maps for use by emergency response organizations, local, county, State and Federal Government agencies, public and private companies and organizations, the media, and the general public.

Users should be aware of the following specific limitations:

- ShakeMaps are automatic computer generated maps that have not necessarily been checked by human oversight. Because the input data is raw and unchecked, the maps may contain errors. The maps are preliminary in nature and will be updated as data arrives from distributed sources.

- Interpolation, contouring, and color-coding can be misleading because data gaps may exist. Caution should be used in deciding which features in the contour patterns are required by the data. Ground-motions and intensities can vary greatly over small distances, so these maps are only approximate; at small scales and away from data points, they may be unreliable.

- The instrumental intensity map is derived from ground-motions recorded by seismographs and represents Modified Mercalli Intensities (MMI) that are likely to have been associated with the ground-motions. Unlike conventional MMI, the estimated intensities are not based directly on observations of earthquake effects on people or structures.

- Locations within the same intensity area will not necessarily experience the same level of damage because damage depends heavily on the type of structure, the nature of the construction, and the details of the ground-motion at that site. For these reasons, more or less damage than described in the MMI scale may occur.

- Large earthquakes can generate very long duration and long period ground-motions that can cause damage at great distances from the epicenter; although the intensity estimated from the ground-motions may be small, significant effects to large structures (bridges, tall buildings, storage tanks) may be notable.

ShakeMap should be regarded as a work in progress. Additional improvements for rapidly and accurately depicting the distribution and intensity of shaking are in progress, and improvements and additions are underway. Further deployment of seismic instrumentation will also lead to significant improvements in the accuracy of the depiction of shaking. To assist us in further improving ShakeMap, users and researchers are invited to submit comments on methodological, software, or presentation issues via the comment form on the ShakeMap World Wide Web homepage at:

http://earthquake.usgs.gov/shakemap

ACKNOWLEDGMENTS

Many contributions in a variety of forms have greatly helped in the development, implementation, and use of ShakeMap. ShakeMap is one important end-product of a very sophisticated seismic network. It can only be produced within the context of a robust, real-time seismic operation. Credit is given to all involved with the regional and national networks in the United States.

Much of the early conceptual development of ShakeMap benefited greatly from discussions with Professors Kanamori and Heaton at Caltech. Both the TriNet Steering and Advisory Committees also provided ongoing oversight and feedback in the early years of TriNet. Discussions with many colleagues, including W. Savage, K. Campbell, R. Nigbor, and M. Petersen, provided additional guidance. Early trips to the Japanese Meteorological Agency (JMA), and in particular discussions with Keiji Doi, were very helpful.

In implementation, Doug Given (USGS) and Phil Maechling and Egill Hauksson (Caltech) were instrumental on the network side of the operation. Engineering-strong-motion and technical advice as well as perspectives from Tony Shakal of the CGS is greatly appreciated. Craig Scrivner, then at the California Department of Mines and Geology (CDMG), contributed greatly to the initial ShakeMap software development.

At regional network centers, Kris Pankow (University of Utah), Steve Malone (University of Washington), Kuo-wan Lin (CGS), Dan McNamara (USGS, Golden), Douglas Dreger, Peter Lombard, and Lind Gee (U.C. Berkeley), Glenn Biasi (University of Nevada, Reno), and Howard Bundock, David Oppenheimer, and Jack Boatwright (USGS, Menlo Park) all played a critical role in system testing, providing feedback, and improving the ShakeMap software. In addition, a number of other people assisted the above colleagues in the regional ShakeMap implementation and operation. Ned Field at the USGS in Pasadena has been very helpful in software calibration and validation and overall advice.

ShakeMap Web pages survived substantial traffic spikes due to the ingenuity and vigilance of Stan Schwarz (USGS, Pasadena). Aesthetic improvements and integration of the ShakeMap Web pages into the USGS Earthquake Hazards Team Web Page standard templates were guided by Lisa Wald (USGS, Golden).

In interfacing with HAZUS with we wish to thank Douglas Huls, Dave Kehrlein, and Lisa Christiansen of the California Office of Emergency Services, Jawhar Bouabid at Durham Technology, and Charles Kircher of Charlie Kircher Assoc. Phil Naecker, Steve Cain, and David Burke of Gatekeeper Systems, Inc., have been enthusiastic and supportive in their development of ShakeCast.

We received extremely important feedback regarding the user interface from participants through a number of meetings and workshops in California for scientific and engineering perspectives, as well as for a very wide variety of users' perspectives. These workshops were usually organized by James Goltz and Margaret Vinci. In addition, ongoing feedback has always been abundant and provides critical advice and ideas that seeds ongoing, iterative improvements to the ShakeMap system.

The manual organization, layout, and document templates were greatly improved by Alicia Hotovec, a summer intern from the Colorado School of Mines. Reviews by Peter Lombard and E.V. Leyendecker improved this manual substantially.

Most of all, we are also extremely grateful for the recognition of the importance of ShakeMap and the ongoing internal and external support for its development at all levels within the U.S. Geological Survey. The support of John Filson, David Applegate, William Leith, Jill McCarthy, Harley Benz, and Woody Savage has been critical.

ANSS ShakeMap Coordinators

David Wald,   *U.S. Geological Survey*, Golden, Colorado, wald@usgs.gov
Bruce Worden,   *U.S. Geological Survey*, Pasadena. cbworden@usgs.gov
Vincent Quitoriano,   *U.S. Geological Survey*, Pasadena, vinceq@usgs.gov
Woody Savage,   *U.S. Geological Survey*, Menlo Park, wusavage@usgs.gov

ShakeMap Regional Coordinators

Southern California:	Bruce Worden, cbworden@usgs.gov
Northern California:	David Oppenheimer, oppenheimer@usgs.gov
	John Boatwright, boat@usgs.gov
	Howard Bundock, bundock@usgs.gov
Utah:	Kris Pankow, pankow@seis.utah.edu
Alaska:	Thomas Murray, tmurray@usgs.gov
	Vincent Quitoriano, vinceq@usgs.gov
Pacific Northwest:	Steve Malone, steve@geophys.washington.edu
Nevada:	Glenn Biasi, glenn@seismo.unr.edu
Central U.S.:	Mitch Withers, mitch@ceri.memphis.edu
Northeast:	Won-Young Kim, wykim@ldeo.columbia.edu
Puerto Rico:	Christa Von Hillenbrandt, christa@midas.uprm.edu

Outreach

James Goltz, *California Governor's Office of Emergency Services,* Pasadena.
Margret Vinci, *California Institute of Technology,* Pasadena.
Lisa Wald, United States Geological Survey, Golden.

1 USERS' GUIDE

ShakeMap originated primarily as an Internet-based system for real-time display. Although the color-coded intensity maps on the Web site are the most visible result of ShakeMap system and constitute the most commonly accessed and downloaded product, they are just one representation of the ShakeMap output. ShakeMap produces grids of acceleration and velocity amplitudes, spectral response values, instrumental intensities, GIS files, and a host of other products for specific users.

In this guide, we describe the basic ShakeMap products and their current and potential uses. First, we provide an overview of the current ShakeMap applications. We then explain the different formats and types of maps available and describe the ShakeMap Web pages. Next, we expand on different automated mechanisms to receive ShakeMap, including new approaches under development, particularly ShakeCast. We also describe Scenario Earthquake ShakeMaps, which provide the basis for pre-earthquake planning and understanding the potential effects of large earthquakes in the future. In each subsection, we try to provide concrete examples of potential uses of each product as well as notable users for each example.

1.1 Introduction

Until recently, the most common information available immediately following a significant earthquake was its magnitude and epicenter. However, the damage pattern is not a simple function of these two parameters alone, and more detailed information must be provided to properly ascertain the situation. For example, for the magnitude-6.7 February 9, 1971, earthquake, the northern San Fernando Valley, California, was the region with the most damage, even though it was more than 15 km from the epicenter. Likewise, areas strongly affected by the 1989 Loma Prieta and 1994 Northridge, California, earthquakes (magnitudes 6.9 and 6.7, respectively) that were either distant from the epicentral region or out of the immediate media limelight were not fully appreciated until long after the initial reports of damage. The full extent of damage from the magnitude-6.9 1995 Kobe, Japan, earthquake was not recognized by the central government in Tokyo until many hours later (e.g., Yamakawa, 1997), seriously delaying rescue and recovery efforts.

A ShakeMap is a representation of ground shaking produced by an earthquake. The information it presents is different from the earthquake magnitude and epicenter that are released after an earthquake because ShakeMap focuses on the ground-shaking produced by the earthquake, rather than the parameters describing the earthquake source. So, although an earthquake has one magnitude and one epicenter, it produces a range of ground shaking levels at sites throughout the region depending on distance from the earthquake, the rock and soil conditions at sites, and variations in the propagation of seismic waves from the earthquake due to complexities in the structure of the Earth's crust.

Part of the strategy for generating rapid-response ground-motion maps was to determine the best format for reliable presentation of the maps given the diverse audience, which includes scientists, businesses, emergency response agencies, media, and the general public. In an effort to simplify and maximize the flow of information to the public, we have developed a means of generating not only peak ground acceleration and velocity maps, but also an instrumentally derived, estimated Modified Mercalli Intensity map. This Instrumental Intensity map makes it easier to relate the recorded ground-motions to the expected felt and damage distribution. We have also further simplified the presentation of the Instrumental Intensity ShakeMap specifically for the resolution and audience of broadcast television to reach the widest audience possible. At the same time, we preserve a full range of utilities of recorded ground-motion data by producing maps of response spectral acceleration, which is not particularly useful to the general public, but which provides fundamental data for loss estimation and engineering assessments.

Although we show several ShakeMap Web page examples in the following documentation, this guide is no substitute for the ShakeMap Web pages, and we recommend having a browser open to those pages while this guide is in hand.

1.2 Current Applications of ShakeMap

Prior to fully describing the array of ShakeMap products and formats, we briefly expand on the most common applications of ShakeMap.

1.2.1 Emergency Response and Loss Estimation

The distribution of shaking in a large earthquake, whether expressed as peak acceleration or intensity, provides responding organizations a significant increment of information beyond magnitude and epicenter. Real-time ground-shaking maps provide an immediate opportunity to assess the scope of an event, that is, to determine what areas were subject to the highest intensities and probable impacts as well as those that received only weak motions and are likely to be undamaged. These maps will certainly find additional utility in supporting decision-making regarding mobilization of resources, mutual aid, damage assessment, and aid to victims

For example, the Hector Mine earthquake of October 16, 1999, provides an important lesson in the use of ShakeMap to assess the scope of the event and to determine the level of mobilization necessary. This earthquake produced ground-motion that was widely felt in the Los Angeles basin and, at least in the immediate aftermath, required an assessment of potential impacts. It was rapidly apparent, based on ShakeMap, that the Hector Mine earthquake was not a disaster and despite an extensive area of strong ground shaking, only a few small desert settlements were affected. Thus, mobilization of a response effort was limited to a small number of companies with infrastructure in the region and brief activations of emergency operations centers in San Bernardino and Riverside Counties and the California Office of Emergency Services (OES), Southern Region.

Quote from a member of a Caltrans County bridge crew, following the 1999 Hector Mine Earthquake:

"I just wanted to say "Thank you" for having your web site made available to everyone on the Internet. As a member of the Caltrans Bridge crew here in San Bernardino county, information on the recent quakes such as the 7.1 we had last weekend was found right here at your site within a few minutes of signing on… I can't tell you how much time and money was saved knowing where to look [for damage] by having this site at our fingertips. Great Work."

Unnecessary response in an effort to fully assess the potential effects of an earthquake, although not as costly as inadequate or misguided response in a real disaster, can be costly as well. Had a magnitude-7 earthquake occurred in urban Los Angeles or another urban area in California, ShakeMap could have been employed to quickly identify the communities and jurisdictions requiring immediate response. To help facilitate the use of ShakeMap in emergency-response, ShakeMap is now provided to organizations with critical emergency response functions automatically through the Internet with electronic "push" technology (see Section 1.5). These organizations and utilities include the State of California OES, the Los Angeles County Office of Emergency Management, Southern California Edison, and the Los Angeles Metropolitan Water District.

ShakeMap ground-motion maps are also customized and formatted into Geographic Information Systems (GIS) shapefiles for direct input into the FEMA's U.S. (HAZUS) loss estimation software. These maps are rapidly and automatically distributed to the California OES for computing HAZUS loss estimates and for coordinating State and Federal response efforts. This is a major improvement in loss-estimation accuracy because actual ground-motion observations are used directly to assess damage rather than relying on simpler estimates based on epicenter and magnitude alone, as was customary.

A ShakeMap-driven calculation of estimated regional losses can provide focus to the mobilization of resources and expedite the local, State, and Federal disaster declaration process, thus initiating the response and recovery machinery of Government. ShakeMap, when overlaid with inventories of critical facilities (e.g., hospitals, police and fire stations, etc.), highways and bridges, and vulnerable structures, provides an important means of prioritizing response. Such response activities include: shelter and mass care, search and rescue, medical emergency services, damage and safety assessment, utility and lifeline restoration, and emergency public information.

In addition to GIS-formatted maps specifically design for HAZUS, we also make shapefiles for more general GIS use. These layers are fundamental as base maps upon which one can overlay a user's infrastructure or inventory. For example, ShakeMaps are also being distributed to regional and State utility providers to enable them to determine areas of their networks that may have sustained damage. Using GIS systems, quick analysis of the situation is possible, and decision-making is greatly facilitated. Insurance, engineering, financial institutions, and others now routinely use these GIS maps for both recent and past earthquakes.

1.2.2 Public Information and Education

The rapid availability of ShakeMap on the Internet combined with the urgent desire for information following a significant earthquake makes this mapping tool a source of emergency public information and education. In instances in which an earthquake receives significant news coverage, the ShakeMap site as well as the Community Internet Intensity Map[1] (which poses the question, "Did you feel it?") receives an enormous increase in Website visitors.

On October 16, 1999, local television stations devoted considerable airtime to the Hector Mine earthquake. During live news briefings, Caltech and USGS scientists employed ShakeMap to discuss the event, invited viewers to visit the ShakeMap Website and posted the Web address prominently above the podium in the media center. By the end of the day, the ShakeMap Website had received more than 300,000 visitors. Even for small events, rapid and reliable earthquake information is important. For instance, on January 13, 2001, when two magnitude-4 events, centered in the northeast San Fernando Valley area of Los Angeles, were followed by local news coverage, Web visits peaked at 233 hits per *second*.

Acknowledging the importance of ShakeMap as a tool for public information and education, we developed a "TV" ShakeMap in cooperation with regional news organizations. This version of ShakeMap represents a substantial simplification of the "official" map that appears on the ShakeMap Website. Based on recommendations of news representatives, acceleration and velocity were omitted from the TV version of ShakeMap. Concern that magnitude and intensity might be confused prompted removal of Roman numerals representing intensity, and intensity was depicted using only the color bar.   Magnitude and location were enlarged and posted at the top of the map.

The ShakeMap for television audiences was developed specifically to encourage broadcast journalists to provide a more accurate depiction of earthquakes in news reports. Prior to ShakeMap, the typical visual representation of an earthquake consisted of a map overlay with the epicenter and radiating concentric rings to represent ground-motion. The patterns of ground-motion are not symmetrical as suggested by these illustrations, and the use of these oversimplified depictions represents an underutilization of available technology by the news media. Use of ShakeMap to discuss an earthquake that has just occurred not only provides a more accurate image of earthquake ground-motion patterns, it also provides important additional information regarding the potential severity of shaking that is useful both to residents of the area impacted and those outside the area who are concerned about friends and family.

ShakeMaps are now reaching a much wider audience through television broadcasting than would be possible through the Internet alone. As an example, a recent magnitude-4.2 earthquake near Valencia on January 28, 2002, which was felt throughout the San Fernando Valley and northern Los Angeles basin, occurred at 9:54 p.m. At least one local news organization *lead* the 10

---

[1] Invites Web visitors (http://earthquake.usgs.gov/shake under "Did You Feel It?") to record their observations on a questionnaire. The data obtained are aggregated to establish a zip-code-based intensity profile for the event (See Wald and others, 1999c, for more details).

o'clock News with a ShakeMap image providing information about the distribution of shaking to millions of viewers only 6 minutes after the shaking.

1.2.3 Earthquake Engineering and Seismological Research

For potentially damaging earthquakes, ShakeMap also produces response spectral acceleration values at three periods (0.3,1.0, and 3 s) for use not only in loss estimation, as mentioned earlier, but also for earthquake engineering analyses. Response spectra for a given location are useful for portraying the potential effects of shaking on particular types of buildings and structures. Following a damaging earthquake, ShakeMaps of spectral response will be key for prioritizing and focusing post-earthquake occupancy and damage inspection by civil engineers.

In addition to providing information on recent events, ShakeMap Web pages provide maps of shaking and ground-motion parameters for past significant earthquakes. Engineers have found these maps helpful in evaluating the maximum and cumulative effects of seismic loading for the life of any particular structure. This is particularly relevant given the recent discovery of the potential damage to column/beam welds in steel buildings following the 1994 Northridge earthquake.

In seismological research, ShakeMap has been proven particularly effective in gaining a quick overview of the effects of geological structure and earthquake rupture processes on the nature of recorded ground-motions. ShakeMaps showing the distribution of recorded peak ground acceleration (PGA) and peak ground velocity (PGV) overlain on regional topography maps allow scientists to gauge the effects of local site amplification because topography is a simple proxy for rock versus deep-basin soil-site conditions. This can lead to more detailed investigations into the nature of the controlling factors in generating localized regions of damaging ground-motions.

1.2.4 Planning and Training: ShakeMap Earthquake Scenarios

In planning and coordinating emergency response, utilities, local government, and other organizations are best served by conducting training exercises based on realistic earthquake situations—ones that they are most likely to face. Scenario earthquakes can fill this role. The ShakeMap system can be used to map ground-motion estimates for earthquake scenarios as well as real data. Scenario maps can be used to examine exposure of structures, lifelines, utilities, and transportation conduits to specific potential earthquakes. ShakeMap automatically includes local effects due to site conditions. The ShakeMap Web pages now have a special section under the *Archives* pages that display selected earthquake scenarios. Additional scenario events will be supplied as they are requested and generated. To contact the ShakeMap Working Group, please use the comment form available on the Web site. The USGS is also planning to make a concerted effort to provide scenario earthquakes online for all regions of the United States.

The U.S. Geological Survey has evaluated the probabilistic hazard from active faults in the United States for the National Seismic Hazard Mapping Project. From these maps it is possible to prioritize the best scenario earthquakes to be used in planning exercises by considering the most likely candidate earthquake fault first, followed by the next likely, and so on. Such an analysis is easily accomplished by hazard disaggregation, in which the contributions of individual earthquakes to the total seismic hazard, their probability of occurrence, and the severity of the ground-motions are ranked. Using the individual components ("disaggregations") of these hazard maps, a user can properly select the appropriate scenarios given their location, regional extent, and specific planning requirements.

Given a selected event, we have developed tools to make it relatively easy to generate a ShakeMap earthquake scenario. First we need to assume a particular fault or fault segment will (or did) rupture over a certain length or segment. We then determine the magnitude of the earthquake based on assumed rupture dimensions. Next, we estimate the ground shaking at all locations in the chosen area around the fault, and then represent these motions visually by producing ShakeMaps. The scenario earthquake ground-motion maps are identical to those made for real earthquakes—with one exception: ShakeMap scenarios are labeled with the word "SCENARIO" prominently displayed to avoid potential confusion with real earthquake occurrences.

At present, ground-motions are estimated using empirical attenuation relationships. We then correct the amplitude at that location based on the local site soil (NEHRP, see Borcherdt, 1994) conditions as we do in the general ShakeMap interpolation scheme. Finiteness is included explicitly, but directivity enters only through the empirical relations. Depending on the level of complexity needed for the scenario, event-specific factors such as directivity and variable slip distribution could also be incorporated in the amplitude estimates fed to ShakeMap. Scenarios are of fundamental interest to scientific audiences interested in the nature of the ground shaking likely experienced in past earthquakes as well as the possible effects due to rupture on known faults in the future. In addition, more detailed and careful analysis of the ground-motion time histories (seismograms) produced by such scenario earthquakes is highly beneficial for earthquake-engineering considerations. Engineers require site-specific ground-motions for detailed structural response analysis of existing structures and future structures designed around specified performance levels. In the near future, we hope these scenarios will also provide synthetic time histories of strong ground-motions that include rupture-directivity effects.

Our ShakeMap earthquake scenarios are an integral part of emergency-response planning. Primary users include city, county, State and Federal Government agencies (e.g., the California Office of Emergency Services, FEMA), and emergency-response planners and managers for utilities, businesses, and other large organizations. Scenarios are particularly useful in planning and exercises when combined with loss-estimation systems such as HAZUS and the Early Post-Earthquake Damage Assessment Tool (EPEDAT), which provide scenario-based estimates of social and economic impacts.

1.3 Maps and Data Products

ShakeMap is fundamentally a geographic product: the spatial representation of the potentially very complex shaking associated with an earthquake. By its complicated nature, we are required to generate numerous maps that portray various aspects of the shaking that are customized for specific uses or audiences. For some uses, it is not the maps but the components that make up the ShakeMaps that are of interest in order to recreate or further customize the maps. In this section we further describe these ShakeMap component products and the variety of maps and formats. Interactive and automatic access to these products is described in sections 2.4.8 and 2.5, respectively.

For each earthquake that warrants generating a ShakeMap, all maps and associated products for that event are available on the earthquake-specific Web pages as described below.

1.3.1 Interpolated Grid File

As described in the *Technical Manual*, the fundamental output product of the ShakeMap processing system is a finely sampled grid of latitude and longitude pairs with associated amplitude values of shaking parameters at each point. These amplitude values are derived by interpolation of a combination of the recorded ground shaking observation and estimated amplitudes at locations that fill in gaps, with consideration of site amplification at all interpolated points. The resulting grid (hereafter, *grid.xyz*) of amplitude values provides the basis for generating color-coded intensity contour maps, for further interpolation to infer shaking at selected locations, and for generating GIS-formatted files for further analyses.

The *grid.xyz* file is an ASCII file contains values that contains X, Y, Z (degrees longitude, degrees latitude, and amplitude, respectively) values of the peak amplitudes at the ShakeMap map grid nodes in the following format:

The first line is a header with:

```
<name/event_ID of event> <mag> <epicentral lat> <epicentral lon> <MMM
DD YYYY> <HH:MM:SS timezone> <W bound> <S bound> <E bound> <N bound>
(Process time: <time>) <Location String>
```

The first 'time' field is the time of the event. 'Process time' is the time this file was last updated. Below is an example of the header for the 1994 Northridge earthquake ShakeMap:

```
Northridge 6.7 34.213 -118.5357 JAN 17 1994 04:30:55 PST -119.1857
33.7775 -117.857 34.6485 (Process Time: Wed Nov 4 17:25:18 1998)
Northridge Earthquake
```

For large or historic earthquakes the "Location String" will usually be the name of the earthquake, otherwise it will be something of the form "12.1 mi. SSW of Carpinteria, CA."

The remaining lines are of the form:

```
<lon> <lat> <pga> <pgv> <ii> <sa03> <sa10> <sa30>
```

where <lon.> is longitude in degrees, <lat> is latitude in degrees, <pga> is peak ground acceleration (PGA) in units of %g, <pgv> is peak ground velocity (PGV) in units of cm/s, <ii> is Instrumental Intensity in decimal intensity values, and <sa> is spectral acceleration in units of %g. Spectral accelerations are provided for periods of 0.3, 1.0, and 3 s, all with 5 percent damping. These are the commonly used and requested periods, and they are fairly standard for a number of loss-estimation algorithms (e.g., HAZUS).

If the grid file name ends with '.zip,' the file has been compressed with the Zip utility and will need to be unzipped before it can be used. The compressed version of the ASCII grid is now our standard.

1.3.2 Grid File Metadata

Because the grid is the fundamental derived product from the ShakeMap processing, it is fully described in an accompanying metadata file following Federal Geographic Data Committee (FGDC) standards for geospatial information. We do not generate metadata for the parametric data, because that is archived by the regional seismic networks. In fact, because all other ShakeMap products are derived from the gird file, it is sufficient to fully characterize only the grid file using the metadata standards.

This metadata file is distributed via the event-specific Web pages for each earthquake on the download page. The metadata are provided in text, HTML, and XML formats.

1.3.3 GIS Products

ShakeMap processing does not occur in a Geographic Information System (GIS), but we post-process the grid file (above) into shapefiles for direct import into GIS. Shapefiles are comprised of three standard associated GIS files:

> *.dbf* = A DBase file with layer attributes
> *.shp* = The file with geographic coordinates
> *.shx* = An index file

In this application, the shapefiles are contour polygons of the peak ground-motion amplitudes in ArcView shapefiles. These contour polygons are actually equal-valued donut-like polygons that sample the contour map at fine enough intervals to accurately represent the surface function. We generate the shapefiles independent of a GIS using a shareware package (shapelib.c), which employs a 4-point method for contouring.

There is an archive of files (three files for each of the mapped parameters) compressed in Zip format.

1.3.3.1 HAZUS'99 Shapefiles and HAZUS-MH Geodatabases

We generate shapefiles that are designed with intervals that are appropriate for use with the Federal Emergency Management Agency's (FEMA) HAZUS software, though they may be imported into any GIS package that can read ArcView shapefiles. Because HAZUS software requires peak ground velocity (PGV) in inches/s, this file may not be suitable for all applications. The contour intervals are $0.04G$ for PGA and the two spectral acceleration parameters (HAZUS only uses the 0.3 and 1. s periods), and 4 inches/s for PGV.

NOTE: HAZUS'99 users can use the hazus.zip shapefiles (see below) directly. However, the 2004 release of HAZUS-MH uses geodatabases, not shapefiles. As of this writing, FEMA has a temporary fix in the form of Visual Basic script that imports ShakeMap shapefiles and exports geodatabases. FEMA has plans to incorporate such a tool directly into HAZUS-MH in the next official release (D. Baush, FEMA, Region VIII, oral commun., 2004).

HAZUS traditionally used the epicenter and magnitude of an earthquake as reported and used empirical relationships to estimate ground-motions over the effected area. These simplified ground estimates would drive the computation of losses to structures and infrastructure, estimates of casualties and displaced households (for more details, see Kircher and others, 1997; FEMA, 1997). With the improvements to seismic systems nationally, particularly in digital strong-motion data acquisition, and the advent of ShakeMap, HAZUS now can directly import a much more accurate description of ground shaking. The improved accuracy of the input to loss-estimation routines can dramatically reduce the uncertainty in loss estimation due to poorly constrained shaking approximations.

The HAZUS GIS files are only generated for events that are larger than (typically) magnitude 5.0. The set of shapefiles for these parameters is an archive of files (three files for each of the mapped parameters) compressed in Zip format (*hazus.zip*) to facilitate file transfer.

An important note on the values of the parameters in the HAZUS shapefiles is that they are empirically corrected from the standard ShakeMap peak ground-motion values to approximate the (geometric) mean values as used for HAZUS loss estimation. HAZUS was calibrated to work with mean ground-motion values (FEMA, 1997). Peak amplitudes are corrected by scaling values down by 15 percent (Campbell, 1997; Joyner, oral commun., 2000).

If you are unfamiliar with using shapefiles to run HAZUS, we have created a brief tutorial in cooperation with the California Office of Emergency Services (OES) that can be downloaded from the ShakeMap Web pages (under ***Products***).

Example Uses and Users: HAZUS loss estimation. HAZUS users can download and import the ShakeMap *hazus.zip* file and data related to estimated losses for the regions. HAZUS output includes numerous GIS maps and tabulated loss estimates, including casualties, building losses, displaced households, amount of debris, and losses to critical facilities lifelines among many other useful estimates. Estimates of direct economic losses from damage are provided. Example users who run HAZUS software include the Federal Emergency Management Agency, California Governor's Office of Emergency Services (OES), and numerous municipalities. Even though HAZUS can take hours to run for a major earthquake, OES is developing tools to separate large regions into multiple areas and operate on them simultaneously with multiprocessor computing platforms. Total losses are aggregated at the end. This greatly reduces the total run time.

1.3.3.2 GIS Shapefile

High-resolution contour polygons for the peak ground-motion parameters are also available as shapefiles intended for use with any GIS software that can read ArcView shapefiles. Note, however, that the peak ground velocity (PGV) contours are in cm/s, and are therefore NOT suitable for HAZUS input.

The contour intervals are 0.04G for peak ground acceleration (PGA) and the three spectral-acceleration parameters (only two of which are used by HAZUS), and 2 cm/s for PGV. The file also includes MMI contour polygons in intervals of 0.2 intensity units. These shapefiles have the same units as the online ShakeMaps.

There is archive of files (three files for each of the mapped parameters) compressed in Zip format called *shape.zip*. The *shape.zip* files is available for all events, but the spectral values are only included for earthquakes of magnitude 5.0 and larger.

Example Uses and Users: Uses include generating GIS poster maps with detailed roadway and urban databases; adding user infrastructure as an overlay on shaking intensity, acceleration, or spectral acceleration. The U.S. Geological Survey uses the shapefiles for generating poster-sized ShakeMaps, including ShakeMap intensity maps into ArcIMS Services (for example, see http://nhss.cr.usgs.gov/) for wide distribution of high-quality map layers, including topography, urbanization, infrastructure, and other geographical databases.

1.4 Web Pages

After triggering, earthquakes are automatically added to the ShakeMap Web page database and are immediately made available through the World Wide Web online interface. Once triggered, the actual processing of the peak acceleration, peak velocity, and intensity maps (including printing and complete Web page generation) takes less than 1 to 2 minutes depending on the size of the earthquakes; larger earthquakes require larger maps to cover the entire shaken area.

The Web maps are interactive. Selection of individual stations on the map initializes a lookup table that provides station information, including station names, coordinates, and the peak ground-motion values recorded on each component. The Web interface thus provides a convenient format for obtaining detailed strong-motion information concerning specific sites. Such information has been long sought following major earthquakes, and now it can be provided rapidly.

The Web site provides access to not only maps of the most recent earthquakes (for instance, a main shock and significant aftershocks) but also all events processed in the past to provide a basis for comparison with recent events. We are also planning on linking the stations to the plots and the database of seismograms so that users can instantly view the entire station recording for that event.

Although ShakeMap is a fundamentally Web-based system, an important goal in the distribution of ShakeMap is to deliver maps rapidly and robustly to critical users independent of Internet load and server capacity or accessibility. For perhaps a majority of users, the Internet will provide a primary and valuable means of access and delivery. For this reason, substantial consideration was put into both local Web page service as well as expanded service through commercial services. These issues will be addressed in Section 2.4.7.

1.4.1 About the Web Pages

The central service site for all ANSS ShakeMap Web pages is through the USGS Earthquake Program Web pages at:

http://earthquake.usgs.gov/shakemap

We have also secured URLs http://www.shakemap.org, which simply redirects for the main page.

ShakeMaps are delivered to servers locally, and in the western, central and eastern regional USGS centers (Menlo Park, CA, Denver, CO, and Reston, VA, respectively) where they are also served. Additionally, these pages are cached and redistributed through a commercial contract with Akamai (see "Capacity" below).

All regional ShakeMaps are served locally but are also delivered to these central servers to avoid local Web traffic congestion after a major regional earthquake. In addition, the California Integrated Seismic Network (CISN), a region of the ANSS, has added further Web server capacity in California via the CISN Web site http://www.cisn.org/.

Direct links (URLs) to regional ShakeMap Web pages (for example in southern California, TriNet at http://www.trinet.org/shakemap) are still populated, but we can only assure sufficient bandwidth through the USGS Earthquake Program pages.

For a new event, all related Web pages are generated as part of the ShakeMap processing systems. In this sense, all maps and Web pages are made, or remade, "on the fly." This includes event-specific pages, the database (*Archives*), and the front home page. Because the actual processing and generation of ShakeMaps takes a minute or two, the first action after notification from the seismic network (triggering a ShakeMap processing run), is to place a "Waiting" Web page online, notifying all potential users that the maps are being processed and to stand by. This action is motivated by the knowledge of thousands of users repeatedly refreshing their browser, looking for the maps. This produced substantial traffic even prior to a new event being posted.

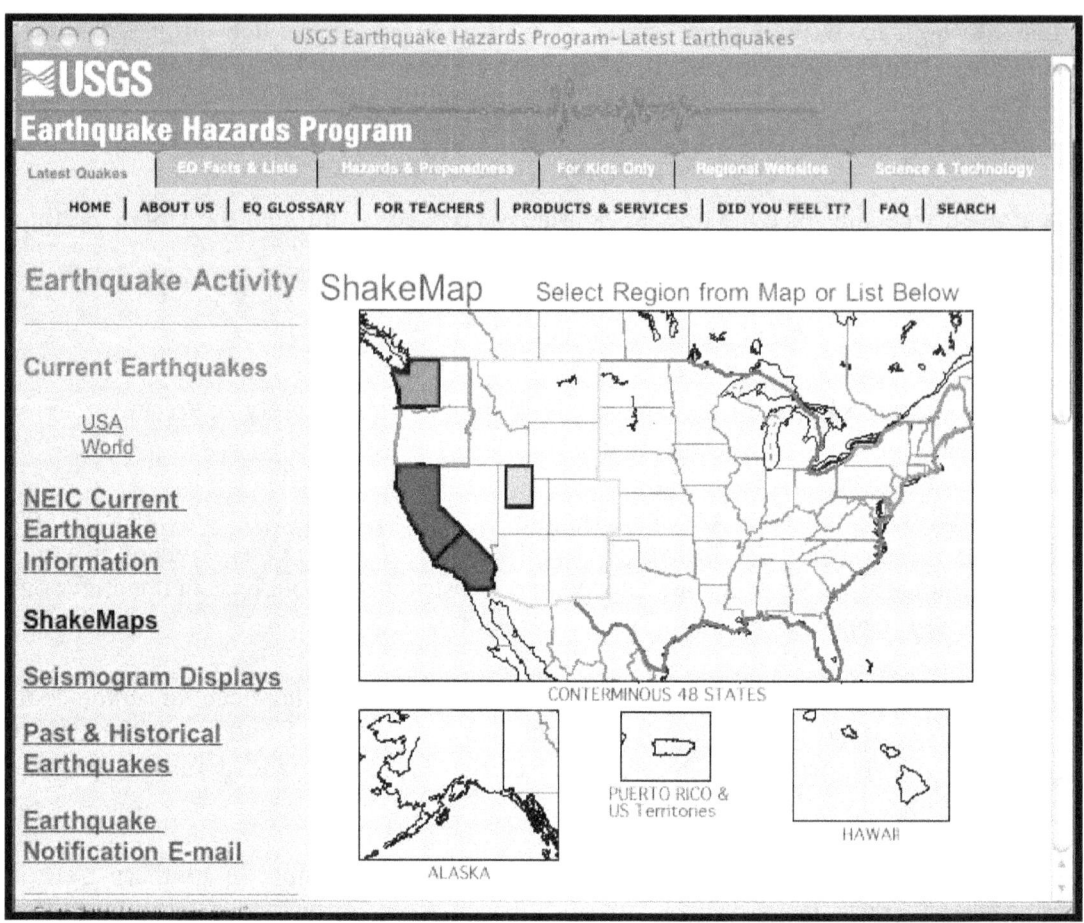

Figure 1.1   National (ANSS) ShakeMap home page. Colored lines indicate continental
U.S. ANSS regions (red, Pacific Northwest; black, California; yellow, Intermountain
West; green, Central U.S.; purple, Northeast). Alaska, Hawaii, and Puerto Rico also
represent separate ANSS regions. Filled, colored areas represent territory covered by
ShakeMap (blue, California; red, western Washington; yellow, Salt Lake City and
environs). Although ShakeMaps are made for earthquakes in these regions, the quality of
the maps is variable and depends on regional seismic-station coverage.

A critical component of the ShakeMap Web pages is that they are static, that is, the content is not
dynamically generated by user-requested actions. Effectively, this means no Web pages are built
based on user requests, and no CPU cycles go toward typical Web user-requested actions that
may normally result in CGI script processing, database searches, interactive forms, etc. In this
way, we can maximize the number of users that we can accommodate. One drawback of this
requirement is that we necessarily limit functionality and sacrifice some desirable map-making
tools that could be allowed with a more regular traffic flow. Recall that our Web pages lie fairly
dormant until an earthquake, at which time Web traffic spikes abruptly. This is discussed further
in Section 1.4.7.

1.4.2 ShakeMap Home Web Page Layout

The basic layout of a regional ShakeMap homepage is shown in Figure 1.1 for northern California. Access is provided to maps for several of the most significant earthquakes in the region, *Archives* of past, significant, and scenario earthquakes, *Related Links*, *Scientific Background*, a *Disclaimer*, and a feedback or *Comment* form. The most significant event is highlighted in red if there are a series of events or a main shock with substantial aftershocks.

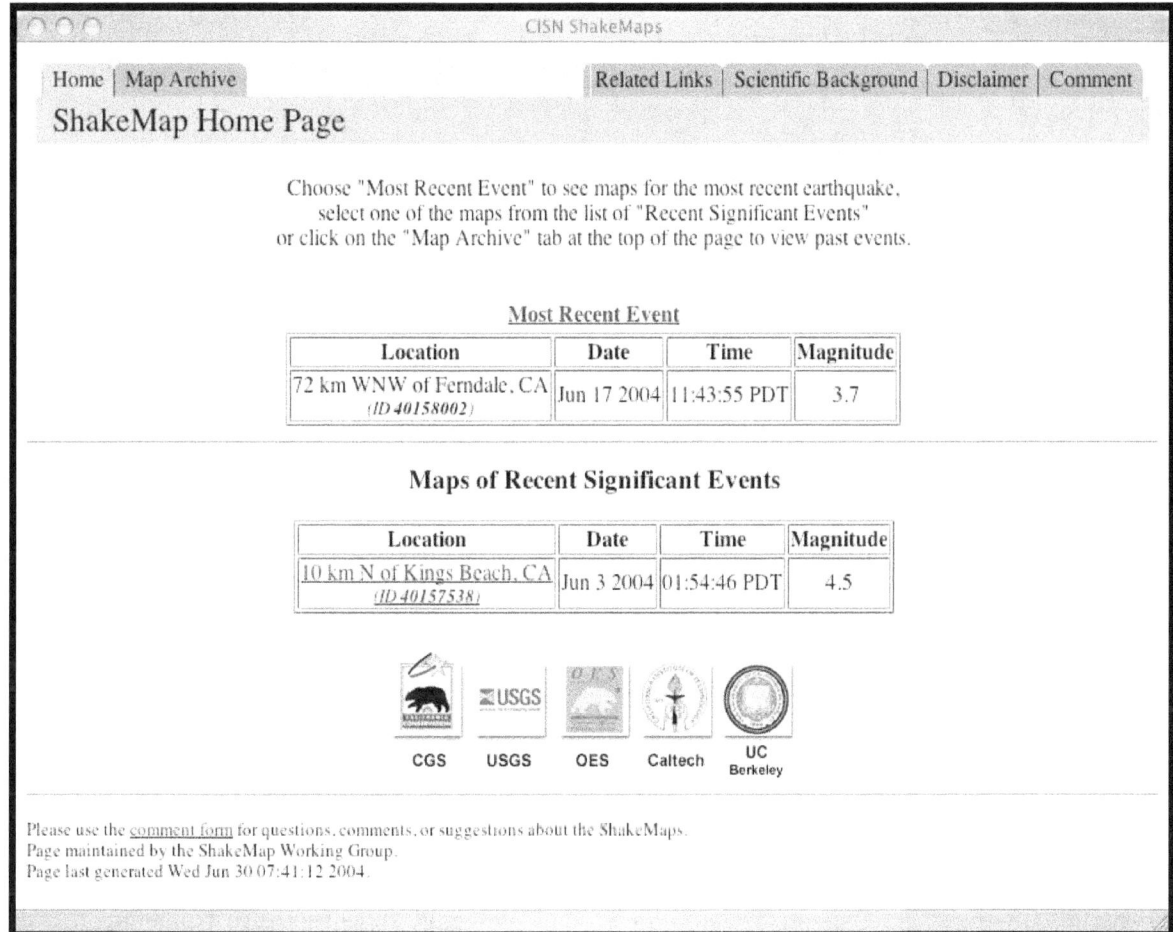

Figure 1.2   Northern California regional ShakeMap home Web page showing recent significant earthquakes in the area. Regional partners in the system are acknowledged with logos on the bottom of the page containing associated URL links.

1.4.3 Individual Event Pages

Selecting any earthquake-specific link brings one to the event-specific page, as shown for example for the December 22, 2003, San Simeon earthquake page shown in Figure 1.3. Whether the event is a recent or past earthquake or a Earthquake Scenario, all subsequent pages are laid out similarly. The only notable difference from event to event is the dependency on magnitude:

spectral acceleration maps are only displayed for events over a configurable threshold, typically magnitude 5.0. For smaller events, these maps are not generated due to lack of need, the reduced signal-to-noise ratio, and to save computational and file-transfer time.

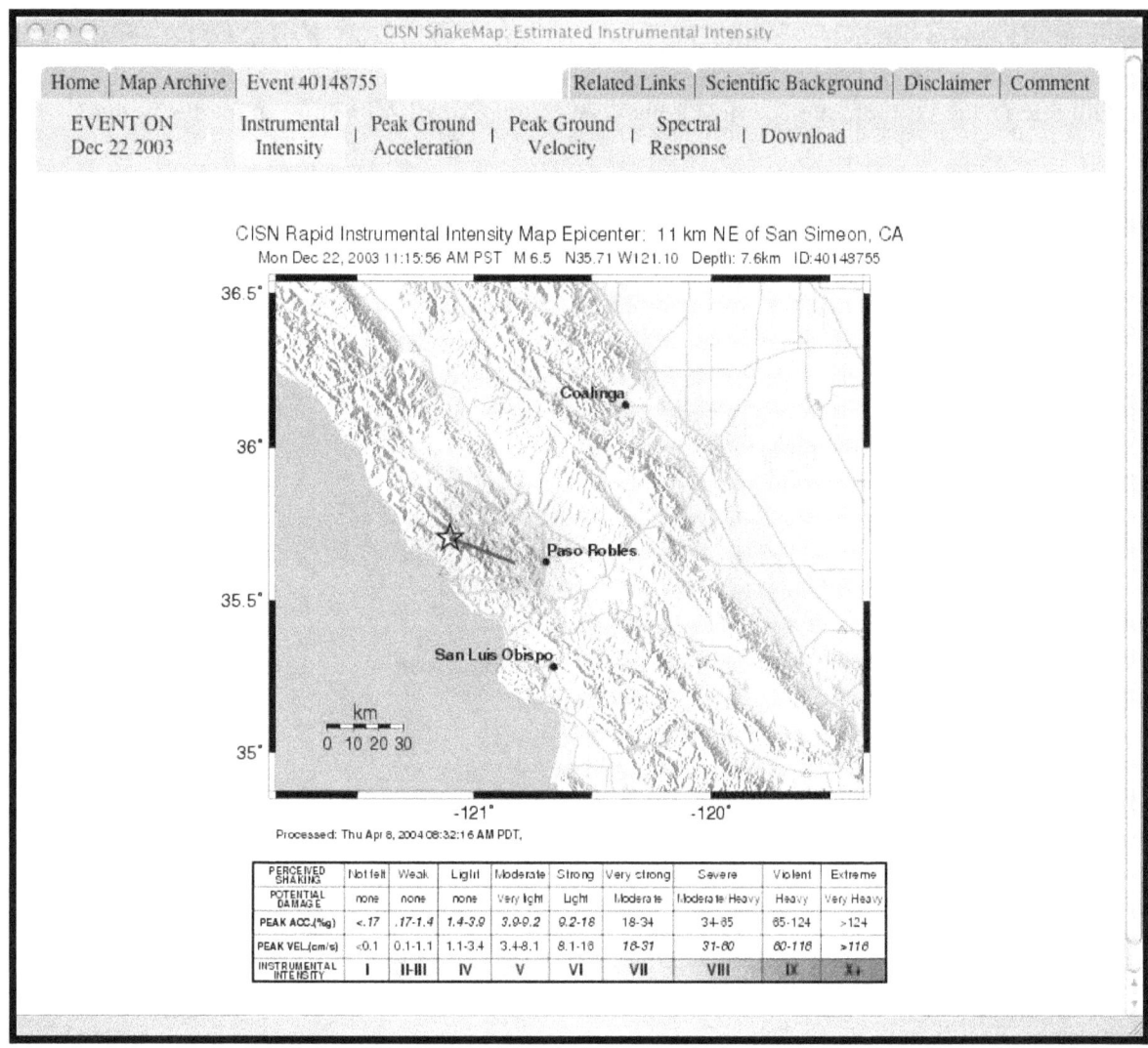

Figure 1.3   Northern California region ShakeMap Web page showing the instrumental intensity map for the magnitude-6.5 San Simeon, California, earthquakes. By default, the intensity map is shown, although peak ground acceleration and velocity as well as spectral response maps are easily accessed via the second row of links above the map.

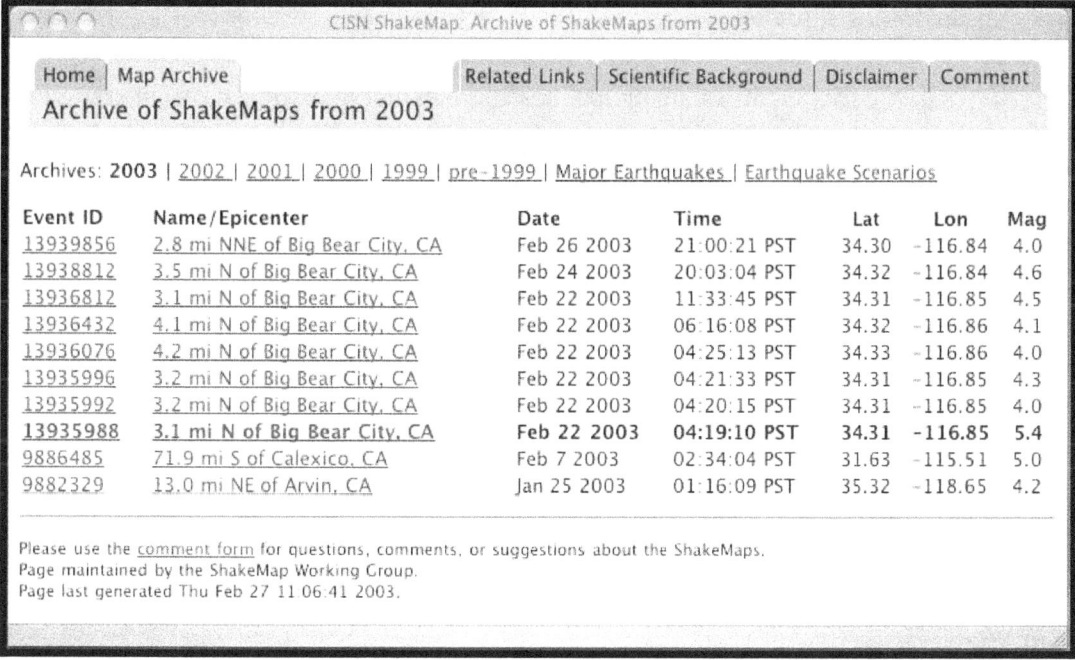

Figure 1.4 Pop-up Web page window showing individual station summary information. This window appears when a station on the ShakeMap is selected with the cursor.

For each individual earthquake, an important tab in addition the maps listed is the *Download* link, which brings up the whole suite of associated maps and products for that earthquake. More information about this page and these products is found below.

1.4.4 Earthquake Archives

An important link on the uppermost row of tabs is the *Map Archive*. Only recent events are linked on the front homepage to insure visitors can find the current earthquake with no effort. However, through the Archives, all past ShakeMap events are listed chronologically, major earthquakes are collated, and a suite of scenario earthquake ShakeMaps are made available.

Figure 1.5   Southern California ShakeMap *Archive* Web Page indicating maps available for the year 2003. Links provide access to other maps for earlier years, major earthquakes in the region, and earthquake scenarios.

1.4.4.1   Recent and Past Events.

A chronological listing of all ShakeMaps made for the region are made via this link. They are listed by year, and then by reverse chronological order from top to bottom. The left-most column in the archive gives the event identification number used by other Web pages that connect the event to the regional seismic network database.

1.4.4.2   Major Earthquakes

Data for the events displayed here may predate the digital networks now operating and contributing to regional ShakeMaps. If a significant earthquake occurred because the beginning of ShakeMap operation in the region, such events are also archived under this heading.

Example Uses and Users: Civil Engineers have used these maps to understand the maximum and cumulative effects of seismic loading for the life of any particular structure for all recent significant earthquakes in Los Angeles (1994 Northridge, 1991 Sierra Madre, 1987 Whittier Narrows, 1971 San Fernando events). This is particularly relevant given the recent discovery of the potential damage to column/beam welds in steel buildings following the 1994 Northridge earthquake. Events with associated damage data have also been extensively used to calibrate loss-estimation software.

1.4.4.3   Scenario Earthquakes

Example Uses and Users: Utilities, municipalities and other large organizations interested in planning response and earthquake drills specific to their area may use the scenario earthquake feature. Earthquake engineers, insurance agencies, and the loss-estimation community also use these events to gauge the impact of individual scenarios on specific inventory or regional exposure.

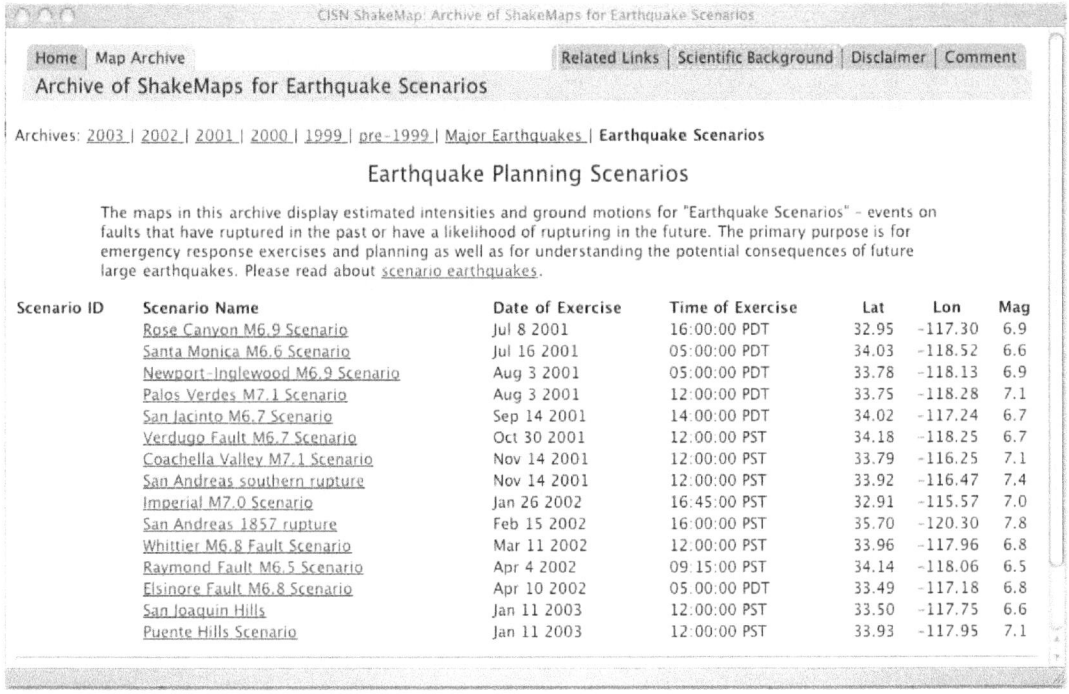

Figure 1.6 Southern California ShakeMap *Scenario Earthquake* Web page. Dates and times of events are either arbitrary or are coordinated to coincide with a particular planning exercise for an earthquake drill as requested by a particular group (usually through the *Comment* form).

1.4.5 Download Pages: A Summary of ShakeMap Products

The *Download* link brings up all associated maps and products for the selected earthquake, whether a recent event, scenario, or major earthquake. Here we summarize the maps, files, data, and information available from this Web page.

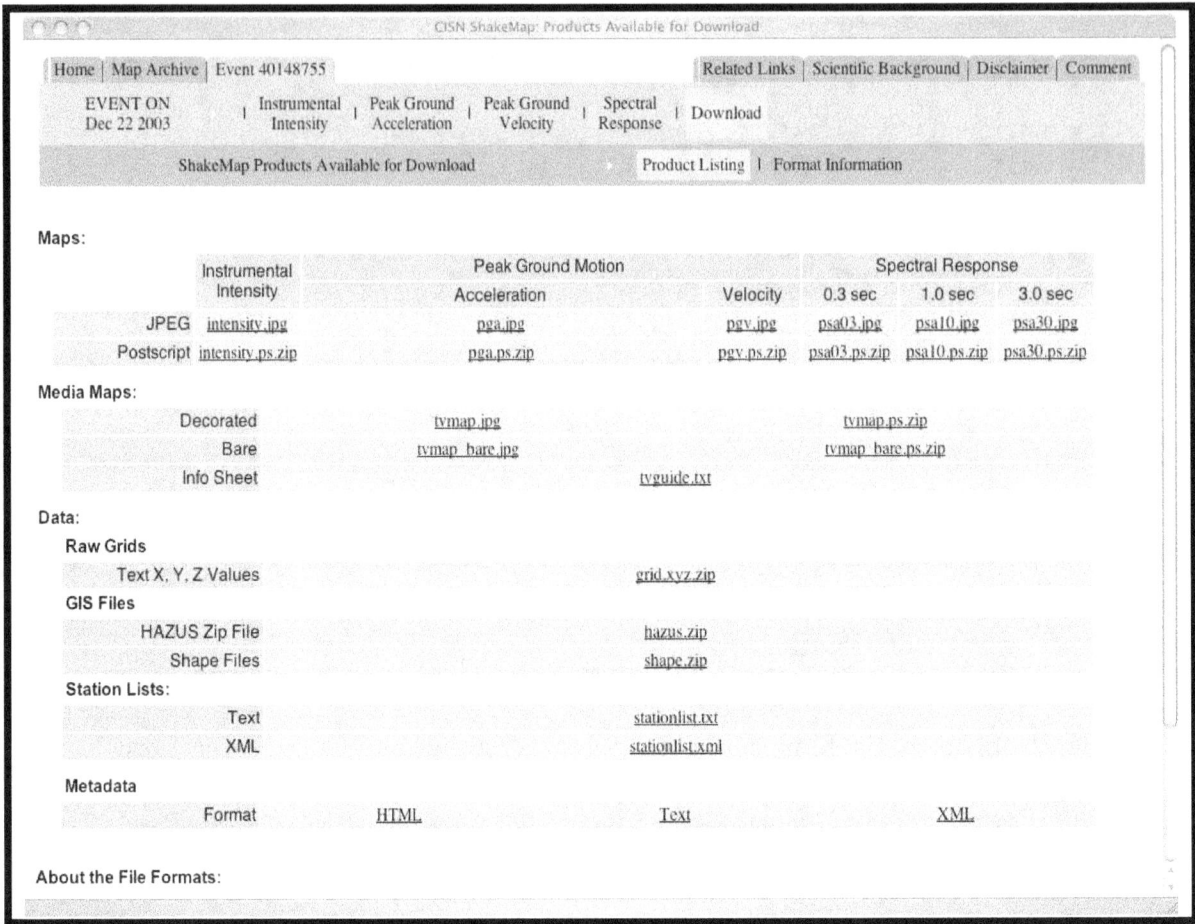

Figure 1.7   ShakeMap *Download* page available for each earthquake.

The products and format descriptions are included in this section. However, note that the link at the bottom of the Download page entitled "***About the File Formats***" provides detailed background for each of the map and product formats available.

Maps:

> ***JPEG***: JPEG (which stands for Joint Photographic Experts Group, the standards body that created it) is a 24-bit, platform-independent image and graphics format. This format can be viewed in any Web browser, and can be manipulated by most image-production applications. The compression scheme is "lossy" though, so multiple generations of editing and saving will degrade the image.

> ***Postscript***: A language to describe graphics independently of the resolution of the output device. Printers with Postscript drivers will rasterize these printer files to high-quality map plots. If the Postscript file name ends with ".zip," the file has been compressed with the Zip utility and will need to be unzipped before it can be used.
> 8-1/2 x 11 Postscript file with map sized to print on 8.5" x 11" paper.

Poster: Postscript file with map sized to print on a poster printer (approximately 32"x28"). This file is only available for large earthquakes.

Media Maps: The Media Maps are simplified versions of the Instrumental Intensity maps (PostScript and JPEG format, see above).

General: Even though the intensity information they contain is exactly the same as that in the other maps, they are packaged in a way that makes them more suitable for broadcast to low-resolution devices, such as TV monitors: roads and borders are thicker; fonts are larger; and the title and intensity scale are simplified.

Decorated: This version shows State borders, map title, simplified intensity scale, and the intensity overlay. This version includes some city names, major freeways, and a distance scale.

*Bare***:** This version shows only State borders, latitude, longitude, and the shaking intensity.

tvguide.txt: This text file is an information sheet intended to supplement the Media Maps. The Info Sheet is a text file that provides basic event information, organizational credits, contact information, and information about earthquake intensities and ShakeMap.

Data:

Station Lists: The earthquake information includes: Event ID, magnitude, date, time, epicenter coordinates and depth. The station information includes name and (or) code, location coordinates, and peak velocity and acceleration values. Stations may be flagged to indicate they were not used in the ShakeMap processing. The types of flags are indicated at the bottom of the list.

*Text***:** A table of earthquake and station parameters, formatted to be read easily by humans.

XML: An XML (Extensible Markup Language) formatted file is also available and is the best option for parsing the information by computer. This is a table of earthquake and station parameters, tagged in XML format for parsing by computer. The DTD defining the structure of the XML flags is incorporated in the file. For more information on XML and XML parsers, see the XML page of the World Wide Web Consortium.

Metadata: ShakeMap produces FGDC-compliant metadata and provides it as text, HTML, and XML on the downloads page. These files are provided to comply with the Federal Geographic Data Committee standards for geospatial metadata. Information regarding the standards can be found at the FGDC Website (http://www.fgdc.gov/metadata/csdgm/). The metadata are provided in text, HTML, and XML formats.

1.4.6 Related Web Pages

1.4.6.1 ShakeMail

Signing up for automatic ShakeMail notification is available through the Related Links tab on the ShakeMap Web pages. Whenever a ShakeMap is made, the user gets notified via email of the creation of the ShakeMap, which is delivered as a JPEG file along with an embedded URL for the event-specific Web pages. Only the initial map is sent via email; updates are not provided with this approach.

1.4.6.2 Add-Ons

ShakeMap produces text strings called "Addons" that are used in conjunction with the ANSS earthquake notification system. With "Addons," all related Web pages that need to know about the availability of these maps received the relevant information and the URL via a system called QDDS, for Quake Data Distribution System (for more information see the QDDS Web pages at ftp://clover.wr.usgs.gov/pub/QDDS/QDDS.html).

1.4.7 Web Server Capacity and Redundancy

Locally (Pasadena and Menlo Park), the ShakeMap Web pages are copied from the local machine generating the maps and pages to the local server. These servers are typically multiprocessor PCs running Free BSD Unix, with a reverse-proxy (Squid) server acting as a memory and request cache to handle the most common requests directly out of main memory. With this approach, the main server has a greatly reduce level (order of magnitude) of requests, expanding the overall capacity of the system. For more information on the Squid Server approach as well as numerous examples of post-earthquake Web traffic spikes, see http://bort.gps.caltech.edu/spikes.

ShakeMaps are delivered to servers locally, and in both east and west cost regional USGS centers (Menlo Park, CA, and Reston, VA) where they are also served. Additionally, these pages are by cached and redistributed through a commercial contract with Akamai (http://www.akamai.com/). Under this contract, capacity is aided by caching and redistribution to over 12,000 servers nationwide.

1.5 Automatic Delivery and Use of ShakeMap

1.5.1 FTP "Push:" Automatic ShakeMap Delivery

We provide a dedicated and automatic delivery mechanism to provide any of the ShakeMap products to critical users employing a standard File Transfer Protocol (FTP) "push."   Most recipients of the ShakeMap push require instant access to the maps and desire automated delivery without having to interactively access and download individual files following a significant earthquake. The FTP push has been very successful in this mode.

This approach requires access through the user's Internet firewall and access to a computer to delivery ShakeMap files. Although robust, this is awkward for some users, and it is now impossible for other potential clients given the more rigorous approach to computer security in recent years. It is often difficult to setup the initial "push" delivery, because this requires substantial coordination with IT security personnel in addition to the communications with the direct ShakeMap users within an organization. Although we have been successful in delivering ShakeMaps with this approach, our daily diagnostic tests reveal various failure modes, making long-term maintenance problematic for ShakeMap operators.

Example Uses and Users: A number of recipients get automatic ShakeMap files and maps delivered via FTP push. Many have developed automated software tools that transfer the files to specific locations, begin loss-estimation routines, and get delivered to in-house GIS databases. These users include the Los Angeles County Office of Emergency Services, Los Angeles Metropolitan Water District, California Governor's Office of Emergency Services, and KNBC Television, among many others.

1.5.2 ShakeCast ("ShakeMap BroadCast")

ShakeCast will allow larger organizations, like Caltrans, and others, to automatically and reliably receive desired ShakeMaps and trigger post-processing tools to initiate an established response protocol. The system will initiate software applications and automatically generate alarms in response to predefined shaking conditions. Currently, USGS "pushes" ShakeMap electronically (using FTP) to utilities and other critical users, but ShakeCast will allow this to be replaced with a subscriber service, providing more robust delivery from redundant ShakeMap generation sites and distributed ShakeCast servers. ShakeCast will also allow organizations to receive and process ShakeMap at multiple divisions within the agency that requires different post-earthquake actions, for instance, Caltrans has post-earthquake responsibilities ranging from bridge inspection and repair to traffic management.

To address these problems, the ShakeCast System is designed to be a simple, reliable, and widely deployable software tool that any modestly capable computer user can install on their computer to receive and make use of customized and personalized earthquake information. We call the system ShakeCast because its purpose is to broadcast ShakeMaps. ShakeCast consists of a receiver component (client) and a transmitter component (server). The information to be disseminated via ShakeCast is the output of the ShakeMap system, which provides early estimates of the severity of shaking during an earthquake and thus is a good tool for estimating the likelihood of damage to structures.

The ShakeCast software will also:

o Automatically download and display maps of the areas affected by an earthquake.

o Automatically receive and process notifications of earthquakes o   Let users define locations (representing structures and facilities) of interest, and set shaking thresholds that will trigger automatic notification

o   Provide users with options for electronic notification (pager, email, personal Web pages, etc.) of events and projected shaking intensity at specified facilities

o   Reliably manage the receipt of updated shaking data from multiple ShakeCast servers distributed around the internet, providing an excellent chance of receiving an uninterrupted and authenticated data feed even after a major event

o   Easily integrated with in-house GIS systems, control systems, utility-outage management systems, and other business systems in organizations

o   Provides a mechanism for continual end-to-end testing of the system, assuring that the system is working properly when it is eventually needed

An overview of the main features of the ShakeCast system being developed is shown in Table 1.1.1 Overview of ShakeCast system features for the client.. ShakeCast allows individuals and facility owners to make widespread and immediate use of the beneficial information already produced by ShakeMap. It takes advantage of the very substantial investment already made in ShakeMap and in the very large seismic monitoring infrastructure behind it. It also provides quantitative metrics on the use of ShakeMaps both before and after an earthquake. These data will then be available for policy decisions on the future direction of the ShakeMap and ShakeCast systems. Finally, ShakeCast should help engage and involve managers and policy makers at a wide variety of institutions (e.g., State transportation departments, municipal governments, emergency responders, utilities, etc.) who are concerned about timely receipt of earthquake shaking data.

| ShakeCast Client (Receiver) Software Features ||
Feature	Description
Multiplatform	Available on PCs and Unix systems
Easy installation and configuration	Installation and basic configuration in less than an hour in most cases
Automated registration	Automatic software registration with ShakeCast broadcast systems, including registration with servers in multiple regions
Integrated quality assurance and testing	The client software will participate in the ShakeCast system's comprehensive end-to-end testing procedures to provide high confidence in proper system function during an earthquake. Broadcast data will be checked for authenticity, correctness, and completeness.
Automated notification	The client software will notify a list of people of earthquake-related events via email, pager, and other mechanisms. Notification can be based on shaking intensity (e.g., "peak ground acceleration at Mom's house greater than 0.3g") using any of the shaking metrics of the current or future ShakeMap system. Users can "sign up" for notification via a Web page on their local ShakeCast system.
Personal Web pages	Provide local ShakeCast users the ability to view shaking data (including maps, events, and alarms) on personalized Web pages served from their local ShakeCast server without each user needing to access the main USGS ShakeMap systems.
Data version support	Revise and reissue notifications as new data arrives. Maintain permanent record of the sequence of notifications issued.
Locations and thresholds database	Maintain local list of locations of interest and notification thresholds.
External program integration	ShakeCast can trigger the execution of external programs for further event and data processing.
Basic GIS tools	Tools for working with GIS format ShakeMap data. Display users own facilities and ShakeMap data in a Web-based map generated locally on the client system.
Simple administration	Web-based configuration and administration interfaces
High-quality documentation	Professionally developed documentation and support materials

Table 1.1.1 Overview of ShakeCast system features for the client.

For more detailed information on ShakeCast, see Wald and others (2003), http://www.shakecast.org, or contact the ShakeMap developers through the ShakeMap Web page *Comment* form.

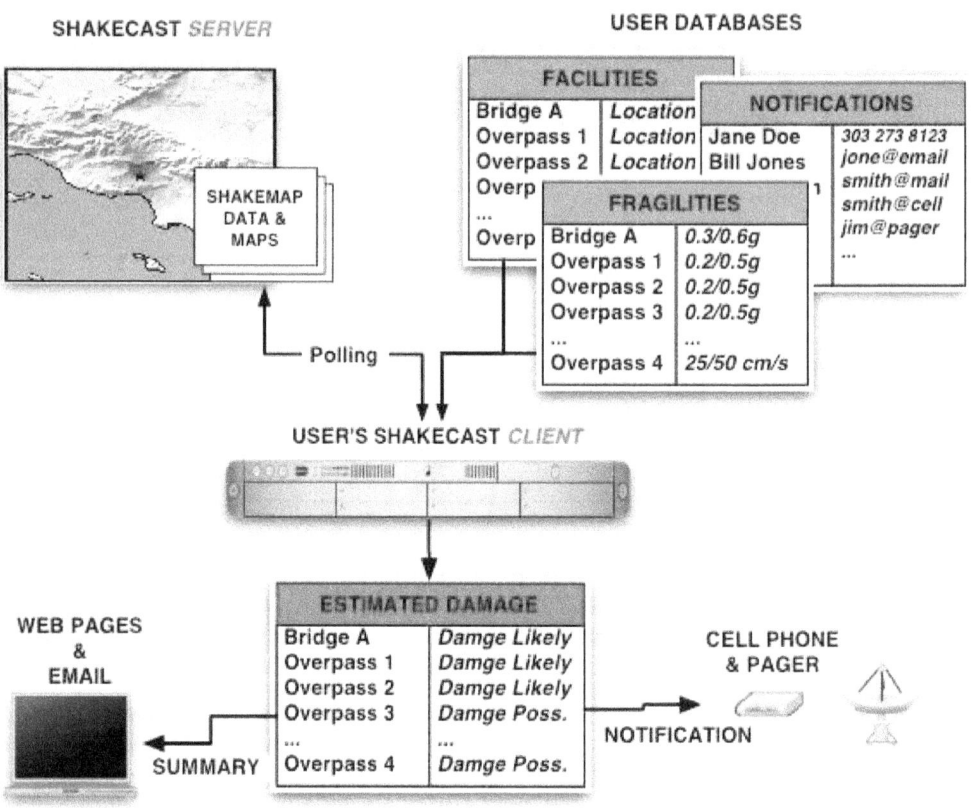

Figure 1.8   Simplified schematic flowchart for the ShakeCast system.

Example Uses and Users: Several ShakeCast users take advance of the build in capacity to determine shaking and potential damage levels at their facilities. Caltrans and Pacific Gas & Electric are testing the system, and FEMA plans to use the system to automatic start up of HAZUS runs to more rapidly estimate overall losses and impact.

1.6 Future Applications of ShakeMap

Ongoing development involves automatically generated, interactive GIS applications for ShakeMap users who are either familiar with or who have expertise in GIS tools and applications. We are implementing both server-side and client-side applications to ensure both diversity of GIS tools and robust access during the immediate post-earthquake time period. Server-side tools allow fully interactive overlays of a variety of ShakeMap parameters and maps with a wide range of regional infrastructure, but their availability is difficult to guarantee in the minutes immediately following a damaging earthquake due extreme demands on the server. In contrast, client-side GIS applications are less versatile, but can be made robust by rapidly and automatically delivering the ShakeMap GIS content (shapefiles) to users.

ShakeMap software has been developed for reliable and robust operation. In addition, the software architecture was designed to be directly portable to other regions of the country. Operating ShakeMap systems now in place cover California as well as the Seattle and Salt Lake City areas. As more seismometers are installed under the Advanced National Seismic System, ShakeMap coverage will be expanded. Regions that will likely come online in the near future include the environs of Memphis, Tennessee, Anchorage, Alaska, Reno, Nevada, and the island of Puerto Rico.

ShakeCast provides many opportunities for automatic and rapid assessment of like impact on distributed facilities for an organization. Efforts are underway to fully develop this system and make it widely available as well as easy to use.

2 TECHNICAL MANUAL

2.1 Introduction

This ShakeMap Technical Manual is meant as the definitive source of information pertaining to the generation of ShakeMaps. The initial description of Wald and others (1999a) is outdated and is superseded by this current report. Technical users of ShakeMap should also consult the User's Guide (Section 1) for additional information pertaining to the format, availability, and the range of ShakeMap-related products available.

Throughout this document, specific parameters that can be configured within the ShakeMap software are indicated in parentheses and are italicized. These configurable parameters are further described in the Software Guide (Section 3).

2.1.1 History and Development

ShakeMap® was originally conceived of by David Wald and designed and implemented by Wald and Vincent Quitoriano in 1996 as soon as a sufficient number of real-time strong motions stations became available by combining the California Seismic Network (Wald and others, 1997) and the newly installed TerraScope stations (Kanamori and others, 1991). Conceptually, we wanted a rapid and automatic, Web-based display of the shaking level at each station on a map generated for each new earthquake, with a location and map scale that would best portray the area shaken.

Due to its utility, the ShakeMap system rapidly evolved during the development, enhancement, and expansion of the TriNet system (Mori and others, 1998 and Hauksson and others, 2002). TriNet was comprised of the U.S. Geological Survey (USGS) Pasadena Field Office, the California Institute of Technology (Caltech) and the California Division of Mines and Geology (CDMG, now the California Geological Survey, CGS) and was funded by the USGS, the California Governor's Office of Emergency Services (OES) through the Federal Emergency Management Agency (FEMA) Hazard Mitigation Grant Program, the California Trade and Commerce Agency, the California Technology Investment Partnership Program and by private-sector contributions.

With the success of the ShakeMap in southern California, a concerted effort was made to enhance the ShakeMap software for distribution to other regional networks around the nation as they gained real-time strong motion capabilities. The original software was then redesigned by Bruce Worden (Caltech, now USGS) and Craig Scrivner (formerly CDMG). Ongoing software development is under the guidance of Worden and Quitoriano as part of the Advanced National Seismic System (ANSS). As described later, ShakeMaps are being generated in other seismically active areas of the United States where funding has allowed sufficient numbers of near-real-time accelerometers.

Deployment of further ShakeMap systems awaits funding and installation of instruments in other urban areas at risk in the United States.

TriNet funding from FEMA ended at the beginning of 2002, however, TriNet continued under the auspices of the California Integrated Seismic Network (CISN) as a region of the Advanced National Seismic System (ANSS; USGS, 1999). Funding for CISN from the USGS continued and increased, and additional funding was provided by the California OES. CISN Statewide coordination includes the three original TriNet partners as well as the Menlo Park office of the USGS and the Seismological Laboratory at the University of California at Berkeley.

Early considerations included deciding on a limited number of ground-motion parameters that could adequately and accurately provide useful post-earthquake information for a wide range of possible audiences. More information on the development and background on the choice and specific uses of each parameter are given in a later section. In addition to the main ShakeMap use—earthquake response—we have added new capabilities to the ShakeMap system, which allows for earthquake planning and response exercises.

In connection with probabilistic hazard maps, ShakeMaps based on earthquake scenarios can also be used to identify points of exposure in lifelines and major structures and to evaluate emergency response plans. They can also be used as a planning tool to identify shortcomings in the existing seismic network and to clarify where resources should be focused. By producing a wide range of products and maps, ShakeMap is also of value to earthquake engineers and earth scientists, as well as the general public.

2.1.2 Other Systems Worldwide

Systems around the world that rapidly provide post-earthquake maps of ground shaking, in addition to simply providing magnitude and epicentral location, are found in the United States (ShakeMap), Taiwan, and Japan. Installation or development of new seismic systems for this purpose is also underway in Canada, Italy, Turkey, and New Zealand.

The Japanese Meteorological Agency (JMA) has provided instrumental intensities (JMA Intensity) because 1996. Ongoing enhancement of the seismic networks that contribute to JMA Intensity Maps expanded greatly after the devastating 1995 Kobe (M6.9) earthquake and now exceeds 4,500 stations, when those of each Prefecture are counted. The density of the observations alone provides a detailed picture of the shaking distribution, and no interpolation is done as in the generation of ShakeMap in the United States. The JMA Intensity maps are routinely and automatically aired on the national television network (NHK) after significant events. In addition, in collaboration with the National Land Agency (NLA), the JMA instrumental intensities can also be used for rapid loss estimation by combining this shaking information with building, census, and infrastructure inventories and detailed knowledge of the geological conditions. Other systems with yet higher spatial station density are also in place in Japan, including more dense local networks like the 150-station network in the City of Yokohama and a several-thousand station network under development by Tokyo Gas. The Tokyo Gas system, referred to as Seismic Information Gathering Network Alert System (or SIGNAL; Shimizu and Yamazaki, 1998) monitors the Tokyo Gas network with 331 accelerometers that telemeter velocity spectrum intensity values (SI). Based on the SI values, Tokyo Gas can rapidly estimate potential damage to gas pipelines with a GIS that facilitates making gas-service shut-off decisions.

The Central Weather Bureau (CWB) in Taiwan has been producing maps of ground acceleration and associated acceleration-based intensities values very rapidly (<2 minutes) following felt events on the island. This system has been in place because the early 1990s, and was shown to be valuable following the devastating 1999 Chi-Chi, Taiwan, (M7.6) earthquake (Wu and others, 2000). With about 80 real-time stations, and well-calibrated site-amplification factors at 700 additional strong motions sites, the system allows interpolation from the 80 real-time recording sites into a more complete picture of the pattern of shaking (Wu and others, 2001). Users of the ground-motion information include the fire response officials who receive summary pager messages of the intensity values at key populated cities over the entire island of Taiwan. Based on the vast data collected during the Chi-Chi earthquake, Wu and others, (2003) began reporting Instrumental Intensity for domestic earthquakes with their rapid reporting system (RRS) by relating intensity to peak ground velocity similar to what is done in the ShakeMap system.

2.2 ShakeMap Software Overview

ShakeMap is a collection of modules written in PERL. PERL is a powerful, freely available scripting language that runs on all computer platforms. The collection of PERL modules allows the processing to flow in discrete steps that can be run collectively or individually. Within the PERL scripts, other software packages are called, specifically packages that enable the graphics. For instance, maps are made using the Generic Mapping Tool (GMT; Wessel and Smith, 1991) and the Postscript output from GMT is converted to JPEG format using Imagemagick. In the design of ShakeMap, all components are built from freely available, open-source packages.

To enable customization for specific earthquakes or for different regions, each ShakeMap module has an accompanying collection of configuration files. For example, in these files, one assigns the regional boundaries and mapping characteristics to be used by GMT, where and how to transfer the maps, email lists and file delivery lists, and so on. Specific details about the software and configuration files are described in detail in the *Software Guide*.

With recent advances in GIS software and usage, several aspects of the ShakeMap system could be accomplished within GIS applications, but the open-source, freely available nature of GMT combined with PERL scripting tools allows for a flexible and readily available ShakeMap software package. Nonetheless, we do take advantage of GIS for a number of products as later described in the *User's Guide*.

2.3 Recorded Ground-motion Parameters

2.3.1 Data Acquisition

For illustrative purposes, we describe the data acquisition in this section primarily for the seismic system in southern California. Some of the details are specific to this network and its particular flow and processing of seismic data. ShakeMap, however, was developed to deal with multiple types of seismic systems, and in later sections we will describe differences in data acquisition at other regional networks within ANSS.

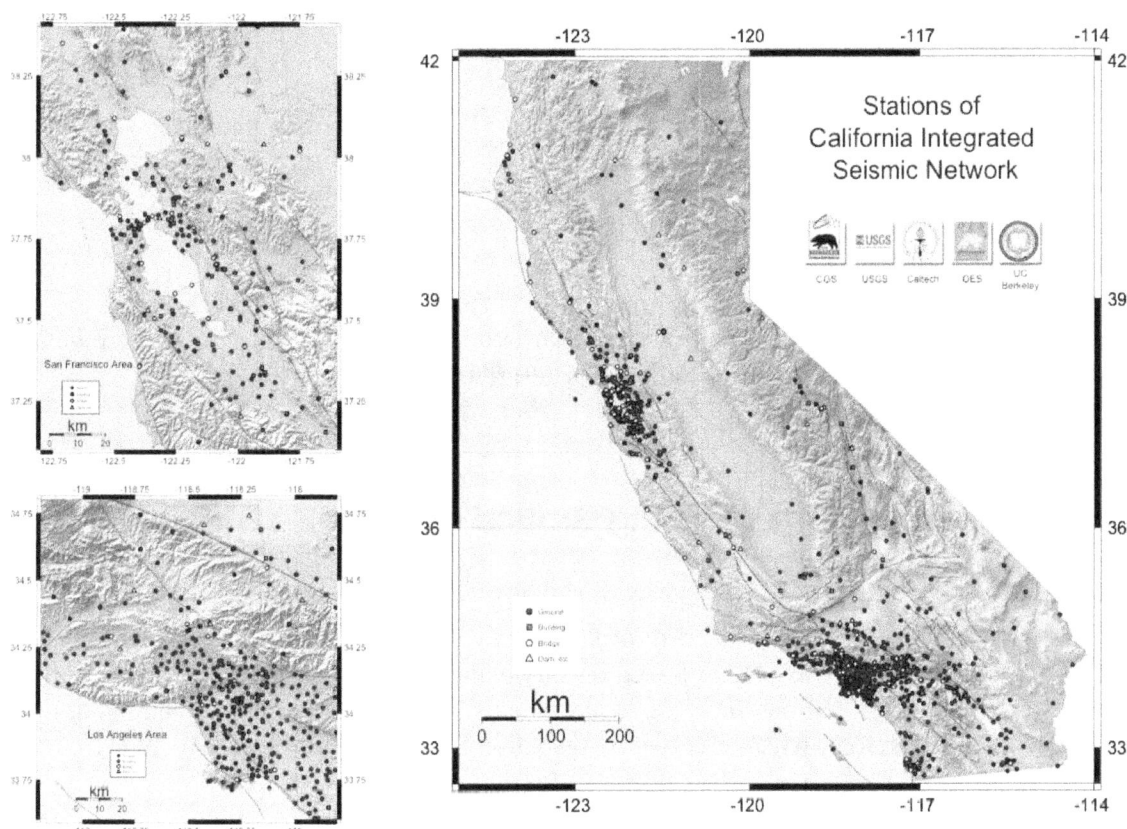

Figure 2.1   Map of the CISN ShakeMap quality seismic station distribution as of July 2004 shown in blue circles. Building strong-motion stations, not used in ShakeMap, are shown as red squares. Figure courtesy of Kuo-Wan Lin.

The seismic station distribution in California is shown in Figure 2.1. Signals from the jointly operated USGS and California Institute of Technology (USGS-Caltech) station are acquired in real time using a variety of digital telemetry methods (see Mori and others, 1998, and Hauksson and others, 2002, for more details). The California Geological Survey, CGS, stations are near real-time, utilizing an automated telephone dial-up procedure (see Shakal *et al*, 1996, 1998). As of March 2002, there are approximately 140 USGS-Caltech real-time stations online and nearly 350 CGS dial-up stations. The USGS National Strong Motion Instrumentation Program (NSMP) also contributes dial-up station parameters within minutes of the earthquake, with nearly 50 stations in southern California alone. Generation of ShakeMap is automatic, triggered by the event associator of the southern California seismic network. Within the first 2 minutes following the earthquake, ground-motion parameters are available from the USGS-Caltech component of the network, and within several minutes most of the important near-source CGS stations contribute. A more complete CGS and NSMP contribution is available approximately within the first 10-15 minutes of the event. Initial maps are made with the real-time component of TriNet as well as any of the dial-sites, and they are updated automatically as more data are acquired.

2.3.2 Derived Parametric Ground-motion Values

Parametric data from the stations include peak ground acceleration (PGA), peak ground velocity (PGV), and peak response spectral acceleration amplitudes (at 0.3 s, 1 s, and 3 s). For the southern California real-time system, values are derived continuously, using recursive, time-domain filtering as described by Kanamori and others (1999). Otherwise parameters are derived from post-processing as described by Shakal and others (1998) and Converse and Brady (1992).

For all maps and products, the motions depicted are peak values as observed; that is, the maximum value observed on the two horizontal components of motion. Many engineers are used to analyses with mean ground-motions, derived from (logarithmic) averaging of the peak values of the two horizontal components, but that is not done for ShakeMap. A more detailed justification for the choice of these parameters is described in Section 1.6.

2.4 Estimating and Interpolating Ground-motions

The overall strategy for the deployment of stations under the ANSS implementation plan relies on dense instrumentation concentrated in urban areas with high seismic hazards (USGS, 1999) and fewer stations in outlying areas. Based on this philosophy, and when fully deployed, maps generated in these urban regions are expected to be most accurate where the population at risk is the greatest, and therefore, where emergency response and recovery efforts will likely be most urgent and complex.

Even so, significant gaps in the observed shaking distribution will likely remain, especially in the transition from urban to suburban to more rural environments, so we have developed algorithms to best describe the shaking in more remote areas by utilizing a variety of seismological tools. In addition to the areas without sufficient instrumentation where we would like to estimate motions to help assess the situation, as a fail-safe backup, it is also useful to have in place the capacity to estimate motions in the event of potential communication dropout from a portion of the network. The same tools are, in fact, beneficial for interpolating between observations (seismic stations) even in densely instrumented portions of the networks.

If there were stations at each of the tens of thousands grid points, then the creation of shaking maps would be relatively simple. Of course stations are not available for all of these grid points, and in many cases grid points may be tens of kilometers from the nearest reporting station. The overall mapping philosophy is to combine information from individual stations, geology (representing site amplification), and ground-motion attenuation for the distance to the epicenter of causative fault to create the best composite map. The procedure should produce reasonable estimates at grid points located far from available data while preserving the detailed shaking information available for regions where there are stations nearby.

Estimating motions where there are few stations and then interpolating the recordings and estimates to a fine grid for mapping and contouring requires several steps. The first stage is to create a coarse, uniformly spaced grid of "phantom stations" using an empirical attenuation relationship that depends on event magnitude and distance (usually epicentral, but may depend on fault finiteness or type of attenuation). These phantom stations are used to estimate shaking in areas far away from reporting stations as if they were recorded on rock site conditions. Those estimates, combined with real stations (also first corrected to approximate rock site conditions), are then interpolated onto a fine-scale grid representing rock motions. The amplitudes at these fine grid stations are then scaled up based on site conditions and are then finally mapped to produce the final ShakeMap product. Each of these steps is described in more detail below.

2.4.1 Phantom Station Grid

We first create a coarse, uniformly spaced grid of "phantom" stations. The choice of phantom stations is fully configurable, but the location and spacing is fixed for each region and the default spacing is usually 30 km. Peak ground-motions are assigned to each coarse grid point using an event-specific, bias-corrected, empirical attenuation relationship based on the magnitude and distance to each grid point (see next section). The bias correction is discussed in a later section. Initially, the distance term defaults to epicentral distance, but in updated maps, we use distance appropriate for the attenuation relationship employed once the fault dimensions can be ascertained (see Section 1.4.4). For Boore and others (1997), which is used in California, this distance is measured from the phantom station to the surface projection of the fault, or simply the fault trace for vertical strike-slip ruptures.

Only those phantom stations farther than a specified distance (default 15 km) from any seismic stations are retained. Likewise, the peak values at the location of the epicenter itself are only used if there are no nearby stations (<10 km). The choices of these two limiting values (pthresh and cthresh, respectively) are configurable. An example of the use of the coarsely gridded, empirically estimated phantom stations is shown in Figure 2.1. Light circles indicate locations of phantom stations. Note that, near the observed strong-motion stations, phantom sites are rejected, allowing the data to control the solution where they exist. For the Northridge earthquake, there is sufficient data in the near-source area that phantom stations mainly fill in gaps, mostly on the outskirts of the map that are at lower ground-motion levels. All other predicted values in this case are superseded by recorded amplitudes. Out at greater distances, however, more phantom stations do contribute to the solution, and they insure that the ground-motion maps remain well behaved and bounded at the edges.

2.4.2 Empirical Ground-motion Equations

The peak ground-motion values for the phantom stations are predicted using an empirical attenuation relation on base rock. Because ShakeMap is run in ANSS regions with varying distance attenuation properties, the choice of attenuation relationships is configurable and expandable. The following table summarizes the available relations, that are used for current regions and for scenario events:

Boore and others (1997), PGV from by Newmark & Hall (1982)	So. California, default regression
Boatwright and others (2003)	No. California, default regression
Atkinson and Boore (2002)	Scenarios only (Cascadia region)
Somerville (1997)	Scenarios only (directivity effects)
Youngs and others (1997)	Washington and Alaska (depth at least 41 km)
ShakeMap Small Regression	All regions (M<5.3)

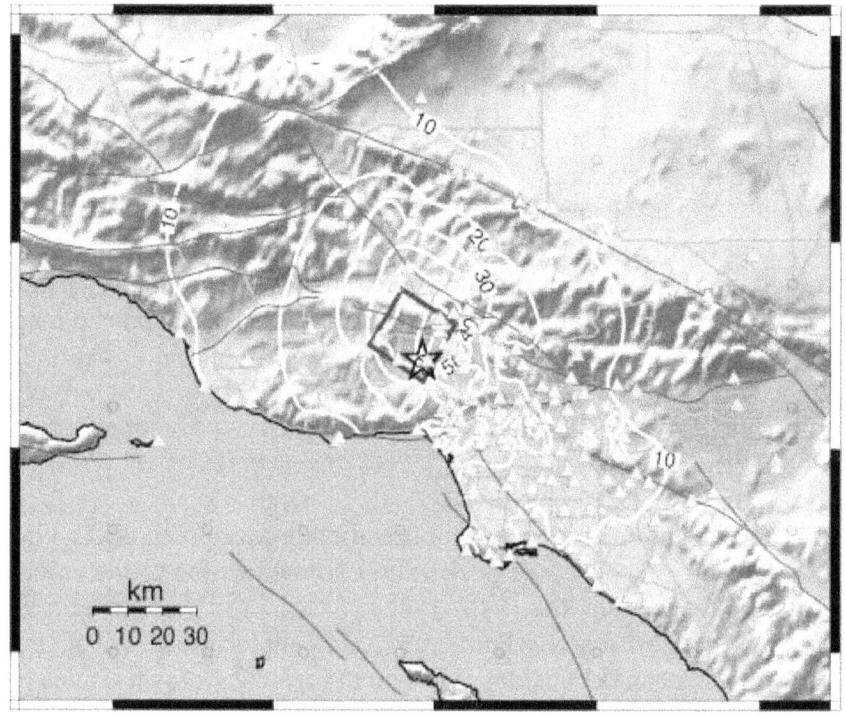

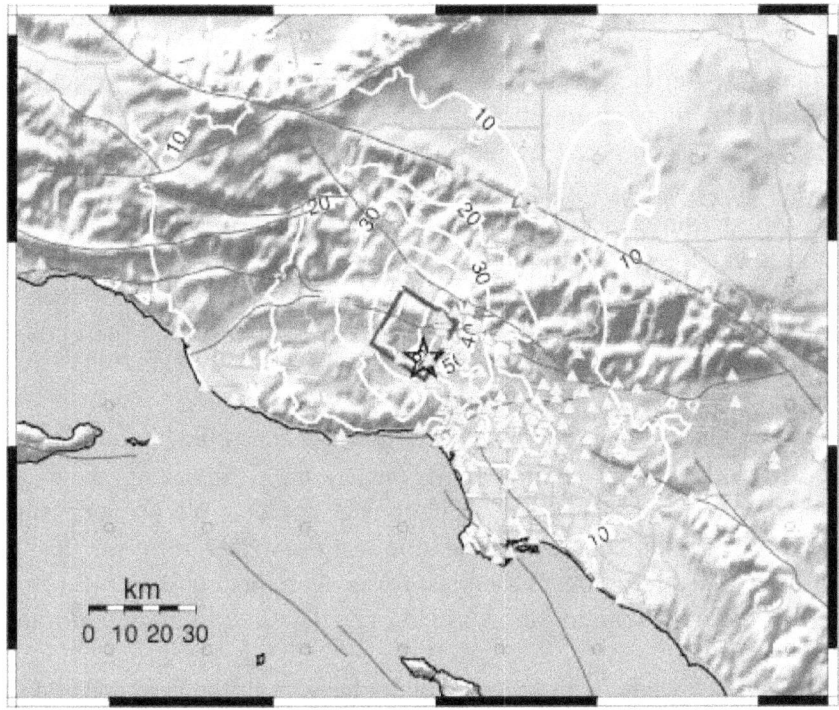

Figure 2.2   Peak acceleration contour ShakeMap for the 1994 Northridge earthquake. Triangles represent stations (pre-TriNet/CISN). The dark-gray-lined polygon is the surface projection of the fault plane from Wald and others (1996). The epicenter is shown with a star; red lines depict faults, light-gray lines show major roadways. Light unfilled circles show locations of empirically predicted "phantom" stations (see text for details). A (top): Without site corrections; B (bottom): With site corrections. Further details for each regression can be found in Appendix A. For this prediction step the baseline 'rock' or 'hard soil' value is used in the attenuation relation. ShakeMap can choose a regression based on event magnitude and depth (when available). The selection rules can be preset for each region. For example, the Southern California ShakeMap uses the Boore and others (1997) regression for events greater than M5.3, and the ShakeMap Small Regression for smaller events.

The predicted values are used to create a 'rock grid' along with site-corrected data from input stations (see Section 1.4.3).

2.4.2.1 Bias Correction

Because we do not typically know the mechanisms of the event at the time ShakeMap is first run, the attenuation relations we use are averages of events of varying mechanisms. Additionally, we are not guaranteed that the initial earthquake magnitude is completely accurate. In addition, because similar magnitude events can have considerable scatter in average ground-motion values, the well documented, so-called inter-event variability (e.g., Boore *et al*, 1997). As expected, this scatter can be considerably different depending on the ground-motion parameter because the dominant period of the parameter in question can be very earthquake dependant.

To overcome these deficiencies, we compute a bias factor for each parameter, by which the predicted ground-motions are multiplied to bring them in line with the recorded data for that event. This factor is computed by minimizing the difference between the data values at the seismic stations and the estimated values at those locations. (In order to remove the effect of site conditions, the station data are first reduced to bedrock values. See Section 1.4.3.) The minimization is in either a least-squares sense or an absolute-deviation sense. Because there is naturally a lot of scatter in seismic data, the absolute deviation (i.e., L1 norm) seems better than an L2 norm and, in fact, has proved to be so in practice, though the choice of norms is also configurable.

In computing the bias we select the distance (in kilometers) beyond which seismic stations will be excluded from the bias calculation (*bias_max_range*); this helps to insure that the bias is computed using the (hopefully) more accurate near-source. We use a default value of 120 km. We also set the minimum number of seismic stations (*bias_min_stations*) within the search radius that are required to compute the bias; fewer than this number will result in the bias being set to 1.0 and a warning message being issued. The default minimum is 6 stations.

For large-magnitude events, with accompanying large fault lengths, it is risky to compute a bias automatically because it will necessarily require the use of an epicentral distance for the initial source-to-station distance calculation. For an extended rupture, the actual distance to many near-fault stations will be much less than the epicentral distance (imagine a great, 400-km-long San Andreas rupture). A bias computed with an assumption of epicentral distance under these conditions will incorrectly overpredict estimated ground-motions. From various tests and experience, the earthquake magnitude above which the bias calculation is not performed (*bias_max_mag*) is given a default value of 7.0. As a side note, this same issue applies to the magnitude calculation; even local energy magnitude will suffer from this distance bias if fault finiteness is not automatically and adequately taken into account.

Finally, we need to be concerned about possible instability in the bias calculation due to bad stations or inadequate representation and some distances. For this reason, the maximum value that the bias is allowed to take (*bias_max_bias*), that is, the maximum factor by which all estimates are multiplied is set to a default value of 4.0. This parameter also sets the minimum bias, which is (1.0 / *bias_max_bias*).

2.4.2.2 Automatically and Manually Removing Outliers

Occasionally, bad data makes it through the system. Normally, with digital telemetry and data processing, clipped data are suitably flagged, but a number of unknown or degenerate cases may occur in which data may be incorrect. We provide two complimentary options. First, we provide a manual flag that removes data supplied from suspected stations. This must be done in advance. Secondly, we cull suspected data by computing the level above and below which data from any station is considered to be an "outlier." We employ the statistics derived for the attenuation relations and specify how many standard deviations define an outlier

(*outlier_deviation _level*). This 'level' can be any positive float, and the default is 3 standard deviations.

We also specify a magnitude above which the automatic flagging of outliers will no longer take place (*outlier_max_mag*) automatically. The purpose of this parameter is to prevent valid data from being flagged because a long fault rupture might cause stations far from the epicenter, but close to the rupture, to show very high amplitudes; the default maximum magnitude is 7.0. The flags vary depending on the reason the station was flagged. Options are listed in the table below.

Station Flagging Codes

Code	Description
M	Manually
O	Outlier
G	Glitch
I	Incomplete trace
N	Not in list of known stations

To automatically or manually force removal of data from suspected stations, rather than simply remove data from the input data files, we specify which stations and components should be flagged in the *flagged_stations.txt* file. The cutoff mentioned above (*outlier_max_mag*) will have no effect on manually flagged stations. Likewise, the manually flagged stations always supersede any automatic flagging introduced. We find it critical that any data removed be so noted, otherwise astute analysts will simply return the suspected data to the input. It is also useful to see that a particular station is flagged (and why) when analyzing the maps. Stations and individual components can be selectively removed by specifying beginning and ending cutoff dates during which data were known to be problematic. We are now developing routines for quick visual review of ShakeMap outliers that will be available immediately to seismic operators. Currently, the list of flagged stations in every event (both manually and automatically removed) is emailed to a list of operators as part of the ShakeMap run. This allows for a rapid check of station reporting and map quality.

Finally, additional configurable parameters specify the minimum regions above and below the PGA and PGV attenuation relation curves in which data values must be accepted and not flagged as outliers (*pga_accepted_halfwidth* and *pgv_accepted_halfwidth,* respectively*)*. This half width overrides the outlier bounds based on the standard deviation of the regression curve, which may be very narrow, particularly at large distances. That is, there may be cases where the sigma values of the regressions (or multiples thereof) are inappropriate to remove outliers because, at great distances, the absolute amplitude values are very small and the scatter about them is large. The default for both parameters is 0.01.

2.4.3 Site Corrections

Site corrections are used to interpolate from ground-motions recorded on a fairly sparse nonuniformly spaced network of stations to maps showing spatially continuous functions (that is, color-coded intensity or contoured peak ground-motion values). For example, direct interpolation between rock sites surrounding a basin may inadequately represent the true, amplified motion within the basin. Prior to interpolation, we reduce the ground-motion amplitudes to a common reference, in this case "bedrock" motions. Recorded peak ground-motion amplitudes from the stations are reduced to rock site conditions (using a procedure described later) and the observations (corrected to rock) and the coarse phantom stations (computed for rock) are then interpolated at points along a fine rock site grid (currently approximately 1.5-km spacing). Finally, the interpolated rock grid is amplified at each point for local site amplification, and a continuous surface, which is fit to the fine grid, is contoured. The finely interpolated grid has been predefined and so we can preassign a geologically based site classification to each location, allowing faster processing.

2.4.3.1   Site Characterization Map

In California, we use the site-conditions map based on geology and shear wave velocity (Wills and others, 2000) shown in Figure 2.3. The California site condition map extent is that of the State boundary, so the southern boundary coincides with the U.S.A./Mexico border. However, due to the abundance of seismic activity in Imperial Valley and northern Mexico, we have continued the trend of the Imperial Valley and Peninsular Ranges south of the border by approximating the geology based on the topography; classification BC (Figure 2.3) was assigned to sites above 100 m in elevation and CD was assigned to those below 100 m. This results in continuity of our site correction across the international border.

2.4.3.2   Amplification Factors

To obtain site amplification factors based on these NERHP site categories, we use the mean shear-wave velocities assigned to them Wills and others (2000), and then apply the frequency- and amplitude-dependent amplification factors determined by Borcherdt (1994) based on these velocities. Given the mean 30-m shear velocities shown in Figure 2.3, the amplifications can be calculated for short-period (0.1-0.5 s) and mid-period (0.4-2.0 s) ranges from Borcherdt (1994, equations 7a and 7b, respectively) at four ranges of input acceleration levels (see Borcherdt, 1994, table 2). These amplification factors are given in Table 2.1. The amplification for the soil sites decreases with increasing ground-motion levels; the rock units have a less pronounced amplitude dependency (Figure 2.3).

We scale the PGA amplitude with the short-period amplification factors, whereas the PGV values are corrected with the mid-period factors. Response spectral values are scaled by the short-period factors at 0.3 s, and by the mid-period response at 1.0 and 3.0 s. The site correction procedure is applied so that the original data values are returned at each station; hence, the actual recorded motions are preserved in the process and the final contours reflect the observations wherever they exist.

Figure 2.3   California Site Condition Map (Wills and others, 2000) based on geology and
correlated to average shear-wave velocity in the top 30 m.

For the reduction of station amplitudes to rock using the amplification factors, the station shear velocity comes from one of two sources. There is a file ("stavel_file") that lists the stations and the 30-m shear velocity at that site. For each station, if such a value is provided in this file it is used; otherwise, the 30-m shear velocity at the station latitude and longitude is sampled from the nearest point on the geology-based site condition grid.

One implication of using site corrections that depend on both frequency and amplitude (Figure 2.3) is that the site corrections are smaller as amplitudes increase into the nonlinear range. Arguably, this range is for peak accelerations above about 20 %g (e.g., Beresnev and Wen, 1996; Field and others, 1997). Hence, for intensity VII or greater, the site corrections (which are based on the peak velocity, or 1 Hz, correction factors) are relatively small.

It will also be important to delineate both the boundaries of potentially damaging near-source strong motions and also those regions at greater distances from the source, where there may be large site amplification. The frequency and amplitude dependence of site amplification on local site geology (average 30-m depth shear velocity) is still a rapidly evolving area of study. Fortunately, modifications to the amplification factors given in Table 2.1 can easily be implemented in ShakeMap as more data and analyses become available.

Site Amplification Factors

Class	Vel		Short-Period (PGA)				Mid-Period (PGV)			
				150	250	350		150	250	350
B	686		1.00	1.00	1.00	1.00	1.00	1.00	1.00	1.00
BC	724		0.98	0.99	0.99	1.00	0.97	0.97	0.97	0.98
C	464		1.15	1.10	1.04	0.98	1.29	1.26	1.23	1.19
CD	372		1.24	1.17	1.06	0.97	1.49	1.44	1.38	1.32
D	301		1.33	1.23	1.09	0.96	1.71	1.64	1.55	1.45
DE	298		1.34	1.23	1.09	0.96	1.72	1.65	1.56	1.46
E	163		1.65	1.43	1.15	0.93	2.55	2.37	2.14	1.91

Table 2.1   Site Correction Amplification factors. Short-Period (.1 to .5 s) factors from equation 7a, Mid-Period (.4 to 2. s) from equation 7b of Borcherdt (1994). Class is NEHRP letter classification; Vel is velocity (m/s) maximum and PGA is cutoff input PGA in gals.

Note that certain regression relations may use their own site amplification method, which supersedes the default corrections. See Appendix A for details on each relation.

2.4.3.3 Interpolation

Maps are prepared by contouring shaking information interpolated onto a rectangular grid uniformly sampled at a spacing interval of approximately 1.5 km (0.0167 degrees, *input_[x,y]_grid_interval*). To help insure accuracy of the map near the edges we also add padding to the edges for all computations (*mapbuf*, set to a value of 0.1 degrees). We then contour the interpolated, site-corrected PGA, PGV, and response spectral values. The interpolation and contouring is done using tools available with Generic Mapping Tools (GMT, Wessel and Smith, 1991).

First, we use the GMT routine *blockmean*, which reads arbitrarily located (latitude, longitude) points and writes out a mean position and value for every block in the define grid region. In the process, *blockmean* acts a filter to avoid spatial aliasing and remove redundant data. We then pass this grid to the routine *surface*, an adjustable-tension continuous curvature surface gridding algorithm that fits the constraining data exactly (Smith and Wessel, 1990).   Hence, our contouring consists of first finding an adjustable-tension (with configurable interior and boundary tension factor, *surface_tension*; default is 0.9), continuous-curvature surface. Then, the GMT tool *grdcontour* is used to produce contour maps and lines. *Grdcontour* simply reads a 2-D gridded file and produces a contour map by tracing each contour through the grid. Much more detailed descriptions of the algorithms involved with the GMT commands *blockmean* and *surface* at the GMT Web site as well as within their application manual pages (http://gmt.soest.hawaii.edu/).

Despite fitting the data in the derivation of the continuous surface, the grid of values sampled from this surface we produce does not include the exact location of the data, unless by close coincidence. For this reason, the exported fine grid we produce is insufficient for recovering the exact values of the data at the original station locations. However, we tabulate these values and provide them with all maps (See *User's Guide*).   Of course, grid nodes nearby a station will be greatly influenced by the data values at that site. A more detailed discussion of the implications for the accuracy of the resulting ShakeMaps can be found in Section 2.7 (ShakeMap Uncertainty).

In Figure 2.2, we show a map of the recorded peak acceleration distribution (contoured in %g) for the 1994 magnitude 6.7 Northridge earthquake to illustrate the nature of the information generated by ShakeMap and the effects of applying the site correction for a larger earthquake. For Figure 2.2a, we have not yet applied the site correction. The contour pattern is only a reflection of the motions as recorded (not corrected to bedrock); In this particular example, the ground-motion data are from existing analog networks (CDMG, USGS, University of Southern California, Southern California Edison, the Los Angeles Department of Water and Power), not the current CISN digital instrument deployment, which postdates the Northridge earthquake. The station density today is comparable to that for this Northridge example; however, these data were not fully available digitally until months after that event.

Typically, for moderate-to-large events, the pattern of peak ground velocity reflects the pattern of the earthquake faulting geometry, with largest amplitudes in the near-source region and in the direction of rupture directivity. For the Northridge earthquake, rupture updip and toward the north resulted in significant directivity in that direction. Differences between rock and soil sites are apparent, but the overall pattern is more a reflection of the source proximity and rupture process. Even though the site effects are still important (see the tabulated amplification factors in Table 2.1), we expect that site corrections for larger events (which are dominated by strong shaking) are less significant than for the lower shaking levels associated with smaller earthquakes. This is particularly true at higher frequencies.

The peak acceleration map for the Northridge earthquake, now applying the ShakeMap site correction approach, is shown in Figure 2.2b. The differences between the ground accelerations within the valleys and surrounding mountains become more evident once the site corrections are applied. In addition, originally smooth contours that simply connected remote stations become more complex when intervening geologically based site corrections play a role in determining the interpolated amplitudes.

From these figures it is clear that the site correction has a more dramatic effect where the station coverage is sparse. Where there are sufficient ground-motion data, the recorded amplitudes *define* the site effects, and nearby site corrections are applied with respect to these observations. In areas lacking observations, the amplitude pattern variations primarily reflect the site corrections modifying an otherwise smoothly varying function of amplitude. In this respect, for areas of sparse coverage, we can consider the application of the geology-based site corrections to be adding data (in the form of our knowledge of site amplification) where there is none.

Note that this approach to interpolation presents an interesting dilemma that has yet to be addressed. If empirically derived, frequency-dependant site amplification factors are available for stations, there is currently no way of implementing them in the ShakeMap algorithm. Although presumably more accurate information would be contained in the empirically derived factors than those based generically on idealized site classifications, the combination of better established amplification factors at randomly located stations and those used for the interpolated grid, which are derived from geology-based inferences, may be in conflict. It this case, there would be many instances where a station and its surrounding nearby grid points would require different amplification factors, resulting in a complex pattern that only reflects the disagreement between map-derived and empirically derived site amplification factors. Using empirically derived amplification factors for a finely spaced grid, perhaps using temporary station arrays, would be one approach.

2.4.4 Fault Finiteness

When the geometry and dimensions of the causative fault become available, this information can then be used for refining the predictive aspects of ShakeMap. In particular, the distance to a given point for empirical regression estimates of shaking are then measured to the fault rather than to the epicenter as is done in the initial, immediate post-earthquake maps. For the Boore and others (1997) regression, for example, distance is then measured to the surface projection of the fault rupture.

In practice, any estimate of the rupture dimensions are placed in a simple text file as ordered pairs of latitude and longitude points and the associated fault depth. In the forward ground-motion estimates, distance to the rupture surface is then computed consistent with the distance measure convention of the specific attenuation relationship being employed. This faulting geometry might be constrained by surface observations, known fault locations combined with aftershock distributions, aftershock locations alone, or from finite-fault modeling when it is available rapidly. Currently, as limited by the current generation of attenuation relationships, slip variations, even if well constrained, cannot be accounted for explicitly; only distance to the fault is considered.

However, if a kinematic finite-fault rupture model is available and forward estimates of the peak ground-motions are computed from that model, we can automatically substitute the modeled (numerical) estimates, which then include both slip distribution and rupture timing, for the empirical estimates obtained from the attenuation relation (by replacing the *estimates.xml* file). This provides event-specific constraints on the ground-motions and can potentially provide a significant improvement over a generic attenuation relationship, even though corrected for a event-specific amplitude bias. In California, this approach depends on the regional waveform modeling approach of Dreger (see Dreger and others, 2000) at the University of California, Berkeley. Based on previous experience, the Berkeley system can provide a robust estimate of the faulting geometry and dimensions in the hours immediately following an earthquake.

For a moderate-sized event with an abundance of ground-motion recordings, such as the Northridge earthquake, adding finiteness has very limited effects because both directivity and fault finiteness are accounted for and are well constrained observationally. For more remote events like the 1999 Hector Mine earthquake, which occurred in the sparsely instrumented Mojave Desert, the addition of the rupture dimension makes a noticeable difference in near-fault ground-motions. Logically, this dictates that dense sampling observationally is necessary in highly populated regions where it is critical to rapidly recover the characteristics of the near-source

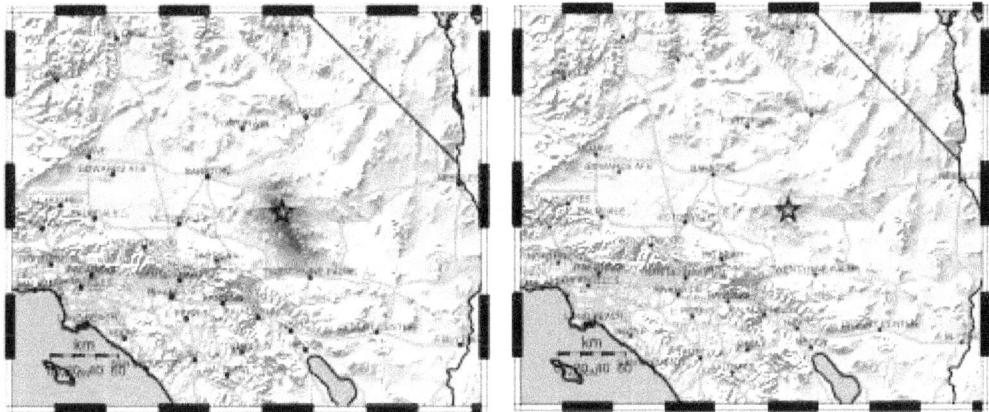

Figure 2.4   Comparison of Hector Mine ShakeMap with fault finiteness (left) and without (right). The map does not change at all in regions with stations, mainly urban areas, but in the remote epicentral region knowledge of the fault dimension changes the picture significantly.

ground-motions. Conversely, despite the significant variations between the Hector Mine map with and without finiteness (Figure 2.4), response and loss estimates based on either map would not vary significantly due to the paucity of inhabitants and associated infrastructure in the near-fault region. In fact, ground-motions for this event were well constrained where significant exposure existed, and these motions did not change with the addition of the faulting dimensions because these locations were observationally controlled. Again, having high station density in urban areas is a stated goal for station deployment within the ANSS (USGS, 1999).

We are currently expanding our capacity to recover source finiteness rapidly by using teleseismic (worldwide) seismic waveforms to independently constrain the source rupture geometry and complexity (see Ji and others, 2003). With such a system, we hope to constrain the rough rupture characteristics with finite fault rupture modeling in the absence of near-fault strong motion data in areas worldwide that are lacking in real-time strong motion networks. Additionally, including surface offset observations, geodetic displacements, regional and local waveforms can be added as they become available.

2.5 Instrumental Intensity

In addition to the PGA, PGV, and spectral response maps, we also map estimates of the ground-motion shaking intensity. Seismic intensity has been traditionally used worldwide as a method for quantifying the shaking pattern and the extent of damage for earthquakes. Though derived prior to the advent of today's modern seismometric instrumentation, seismic intensity still provides a useful means of describing information contained in these recordings. Such simplification is helpful for those users who are unfamiliar with instrumental ground-motion parameters.

That is not to say that instrumentally derived seismic intensity alone is sufficient for loss estimation. In fact, peak velocity and spectral response provide a more physical basis for such analyses. However, for the majority of users, we expect that the intensity map will be more readily interpreted than other maps of ground-motion parameters and will be, therefore, more useful.

2.5.1 Converting from Peak Acceleration and Velocity to Instrumental Intensity

Wald and others (1999b) recently developed regression relationships between Modified Mercalli intensity I_{mm} (Wood and Neumann, 1931, later revised by Richter, 1958) and PGA or PGV specifically for ShakeMap use by comparing the peak ground-motions to observed intensities for eight significant California earthquakes. For the limited range of Modified Mercalli intensities V $\leq I_{mm} \leq$ VIII, Wald and others (1999a) found that for PGA,

$$I_{mm}= 3.66 \log (PGA) - 1.66 \qquad (sigma = 1.08) \qquad (1.1)$$

and for peak velocity (PGV) within the range V $\leq I_{mm} \leq$ IX,

$$I_{mm} = 3.47 \log (PGV) + 2.35 \qquad (sigma = 0.98) \qquad (1.2)$$

Because we are also interested in estimating intensity at lower values, and our current collection of data from historical earthquakes does not provide constraints for lower intensity, we have imposed the following relationship between PGA and I_{mm}:

$$I_{mm} = 2.20 \log (PGA) + 1.00 \qquad\qquad (1.3)$$

This basis for the above relationship comes from correlation of peak ground-motions for recent magnitude 3.5 to 5.0 earthquakes in southern California with intensities derived from voluntary response from Internet users (Wald and others, 1999c) for the same events. We determined that the boundary between "not felt" and "felt" (I_{mm} I and II, respectively) regions corresponds to approximately 1 to 2 cm/s/s, at least for this range of magnitudes. We then assigned the slope such that the curve would intersect the relationship in equation 1 at I_{mm} = V. This relationship may need to be refined as more digital data become available. The corresponding equation for PGV and I_{mm} is:

$$I_{mm} = 2.10 \log (PGV) + 3.40 \qquad\qquad (1.4)$$

By comparing maps of instrumental intensities with I_{mm} for eight significant California earthquakes (see Wald and others, 1999b) we have found that a relationship that follows acceleration for I_{mm} < VII and follows velocity for I_{mm} > VII works fairly well in reproducing the observed I_{mm}. In practice, we compute the I_{mm} from the I_{mm} verses PGA relationship (equations 1.1 and 1.2), and if the intensity value determined from peak acceleration is ≥ VII, we then use the value of I_{mm} derived from the I_{mm} verses PGV relationship (equation 1.2). If the I_{mm} determined from PGA is between V and VII, we weight both the PGA-derived and PGV-derived values, weighted by a factor linearly ramping from 1.0 for PGA at I_{mm} V to 0.0 at I_{mm} VII and vice versa. The switch to PGV for higher intensity insures that spurious high-frequency acceleration spikes will not result in high intensities because the corresponding velocity for such a spike will be low. With our procedure, whereas the large acceleration peak would provide an abnormally high intensity, the much smaller velocity amplitude would provide a more appropriate, lower intensity.

Using peak acceleration to estimate low intensities is intuitively consistent with the notion that lower (<VI) intensities are assigned based on felt accounts, and people are more sensitive to ground acceleration than velocity. Higher intensities are defined by the level of damage; the onset of damage at the intensity VI to VII range is usually characterized by brittle-type failures (masonry walls, chimneys, unreinforced masonry, etc.), which are sensitive to higher frequency accelerations. With more substantial damage (VII and greater), failure begins in more flexible structures, for which peak velocity is more indicative of failure (Hall and others, 1996). This practice is consistent with the recent analysis of Sokolov (1998) in which it was shown that seismic intensities correlate well for rather narrow ranges of Fourier amplitude spectra of ground acceleration, with 0.7-1.0 Hz being most representative of I_{mm} > VIII, whereas the 3-6 Hz range best represents I_{mm} V to VII, and the 7-8 Hz range best correlates with the lowest I_{mm} range. In addition, Boatwright and others (2001) have found that for the Northridge earthquake, PGV and the 3-0.3 Hz averaged spectral velocity are better correlated with intensity (VI and greater) than peak acceleration and their correlation with intensity and peak spectral velocity is strongest at 0.67 Hz.

Figure 2.5 gives the peak ground-motions that correspond to each unit Modified Mercalli intensity value according to our regression of the observed peak ground-motions and intensities for California earthquakes. In assigning integer intensity values using equations 1.1-1.4, the rounding adheres to the convention that, for example, values between 5.50 and 6.49 round to intensity VI. As seen in Figure 2.5, in general a factor of two change in PGA or PGV corresponds approximately to a full step in intensity.

2.5.2 ShakeMap Instrumental Intensity Scale Text Descriptions

Note that the estimated intensity map is derived from ground-motions recorded by accelerographs and represents intensities that are likely to have been associated with the ground-motions. However, unlike conventional intensities, the instrumental intensities are not based on observations of the earthquake effects on people or structures. The terms "perceived shaking" and "potential damage" in the ShakeMap Legend are chosen for this reason; these intensities were not observed, but they are consistent on average with intensities at these ranges of ground-motions recorded in a number of past earthquakes (Wald and others, 1999b). Two-word descriptions of both shaking and damage levels are provided to easily summarize the effects in an area; they were derived with careful consideration of the existing descriptions in the Modified Mercalli descriptions (L. Dengler and J. Dewey, written commun., 1998, 2003).

PERCEIVED SHAKING	Not felt	Weak	Light	Moderate	Strong	Very strong	Severe	Violent	Extreme
POTENTIAL DAMAGE	none	none	none	Very light	Light	Moderate	Moderate/Heavy	Heavy	Very Heavy
PEAK ACC.(%g)	<.17	.17-14	1.4-39	3.9-92	9.2-18	18-34	34-65	65-124	>124
PEAK VEL.(cm/s)	<0.1	0.1-1.1	1.1-3.4	3.4-8.1	8.1-16	16-31	31-60	60-116	>116
INSTRUMENTAL INTENSITY	I	II-III	IV	V	VI	VII	VIII	IX	X+

Figure 2.5   ShakeMap Instrumental Intensity Scale Legend: Color palette, two-word text descriptors, and ranges of peak motions for Instrumental Intensities.

The ShakeMap qualitative descriptions of shaking are intended to be consistent with how people perceive the shaking in earthquakes. The descriptions for intensities up to VII are constrained by the work of Dengler and Dewey (1998) did, in which they compared results of telephone surveys with USGS MMI intensities for the 1994 (Figure 2.6) Northridge earthquake. The ShakeMap descriptions up to intensity VII may be viewed as a rendering of Dengler and Dewey's Figure 7a.

The instrumental intensity map for the Northridge earthquake shares most of the notable features of the Modified Mercalli map prepared by the USGS (Dewey and others, 1995), including the relatively high intensities near Santa Monica and southeast of the epicenter near Sherman Oaks. However, in general, the area of I_{mm} IX on the instrumentally derived intensity map is slightly larger than on the USGS Modified Mercalli intensity map. This reflects the fact that although much of the Santa Susanna mountains, north and northwest of the epicenter, were very strongly shaken, the region is also sparsely populated, hence, observed intensities were not determined there. This is a *fundamental difference* between observed and instrumentally-derived intensities: Instrumental intensities will show high levels of strong shaking, independent of the exposure of populations and buildings; observed intensities only represent intensities where there are structures to damage and people to experience the earthquake.

The ShakeMap descriptions of Shaking begin to lose meaning above VII or VIII. In the Dengler and Dewey study, peoples' perception of shaking began to saturate in the intensity VII -- VIII range, with more than half the people at VII-VIII and above reporting the shaking as "violent" on a scale from "weak" to "violent." In the ShakeMap descriptions, we intensified the descriptions of shaking with increases of intensity above VII, because the evidence from instrumental data is that the shaking is stronger. But we know of no solid evidence that one could discriminate intensities higher than VII on the basis of different individuals' descriptions of perceived shaking alone.

ShakeMap is not unique in describing intensity VI as corresponding to strong shaking. In the 7-point Japanese macroseismic scale, for which intensity 4 is equivalent to MMI VI, intensity 4 is described as "strong." In the European Macroseismic Scale 1998, which is more or less equivalent to the MMI, the bullet description of intensity V is "strong." Higher EMS-98 intensities are given bullet descriptions in terms of the damage they produce, rather than the strength of perceived shaking.

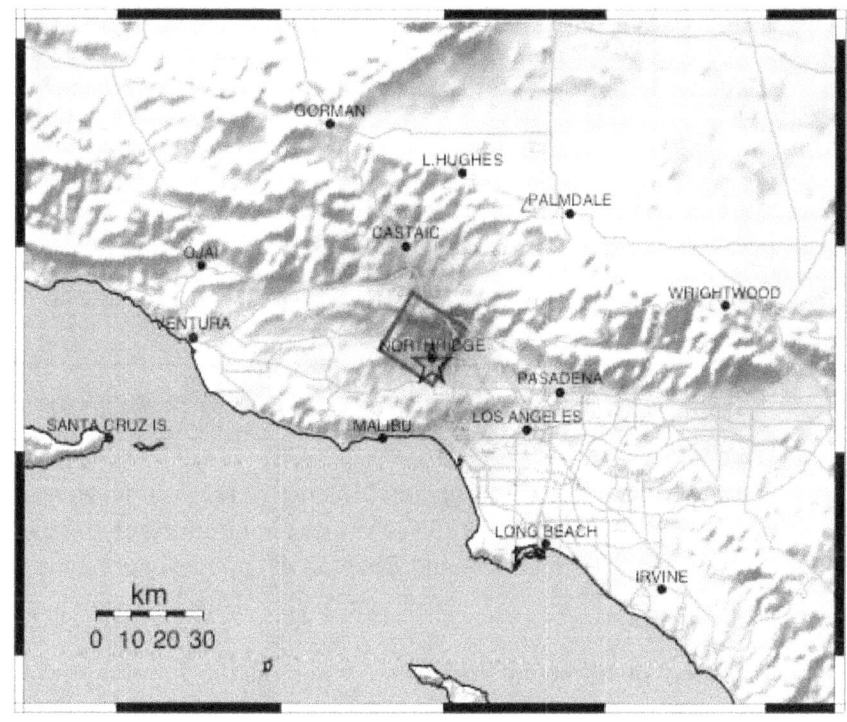

PERCEIVED SHAKING	Not felt	Weak	Light	Moderate	Strong	Very strong	Severe	Violent	Extreme
POTENTIAL DAMAGE	none	none	none	Very light	Light	Moderate	Moderate/Heavy	Heavy	Very Heavy
PEAK ACC.(%g)	<.17	.17-1.4	1.4-3.9	3.9-9.2	9.2-18	18-34	34-65	65-124	>124
PEAK VEL.(cm/s)	<0.1	0.1-1.1	1.1-3.4	3.4-8.1	8.1-16	16-31	31-60	60-116	>116
INSTRUMENTAL INTENSITY	I	II-III	IV	V	VI	VII	VIII	IX	X+

Figure 2.6   Northridge Instrumental Intensity Map. Shaded relief map showing recorded peak instrumental intensity for the magnitude 6.7, 1994 Northridge earthquake. The open star shows the epicenter and the black rectangle depicts the fault surface projection.

2.5.3 Color Palette for the ShakeMap Instrumental Intensity Scale

Color-coding for the Instrumental Intensity map is a standard rainbow palette (see Table 2.2). Such a "cool" to "hot" color scheme is familiar to most and is readily recognizable as it is used as a standard (for example, see USA Today's daily weather temperature maps of the US). Note that we do not feel like intensity II and III can be consistently distinguished from ground-motions alone, so they are grouped together (Figure 2.5). In addition, we saturate intensity X+ with dark red; observed ground-motions alone are not sufficient to warrant any higher intensities given the empirical relationship used does not have any values of intensity greater than IX. In recent years, the USGS has limited observed Modified Mercalli intensities to IX, reserving intensity X for possible future observations (see Dewey and others, 1995, for more details);   no longer do they assign intensity XI and XII.

Intensity	Red	Green	Blue	Intensity	Red	Green	Blue
0	255	255	255	1	255	255	255
1	255	255	255	2	191	204	255
2	191	204	255	3	160	230	255
3	160	230	255	4	128	255	255
4	128	255	255	5	122	255	147
5	122	255	147	6	255	255	0
6	255	255	0	7	255	200	0
7	255	200	0	8	255	145	0
8	255	145	0	9	255	0	0
9	255	0	0	10	200	0	0
10	200	0	0	13	128	0	0

Table 2.2   Color Mapping Table for Instrumental Intensity. This is a portion of the Generic Mapping Tools (GMT) "cpt" file. Color values for intermediate intensities are linearly interpolated from the Red, Green, and Blue (RGB) values in columns 2-4 to columns 6-8.

We drape the color-coded Instrumental Intensity values on the topography to maximize the information available in terms of both geographic location and likely site conditions. Topography does serve as a simple yet effective proxy for examining basin amplification.

By relating recorded peak ground-motions to Modified Mercalli Intensities, we can now generate instrumental intensities within a few minutes of the event. With the color-coding and two-word text descriptors, we can now adequately describe the associated perceived shaking and potential damage consistent with both human and damage assessments of the effects of past earthquakes.

2.6 Discussion of Chosen Map Parameters

2.6.1 Use of Peak Values Rather than Mean

With ShakeMap, we chose to represent peak ground-motions as recorded. We depict the larger of the two horizontal components, rather than as either a vector sum, or as a mean value. The initial choice of peak values was necessitated by the fact that roughly two thirds of the TriNet strong motion data (the CGS data) are delivered as peak values for individual components of motion, that is, as parametric data, not waveforms. This left two options: provide peak values or mean values; determining vector sums of the two horizontal components was not an option because the peak values on each component do not necessarily occur at the same time.

We chose to map peak ground-motion values. Despite the common use of median values in attenuation relations and loss-estimation, we decided that computing and depicting median values, which effectively reduces information and discards the largest values of shaking, was not acceptable. This is particularly true for highly directional, near-fault pulse-like ground-motions, for which peak velocities can be large on one component and small on the other. Mean values for such motions (particularly when determined in log space) can seriously under-represent the largest motion that a building may have experienced, so that option was discarded. What's more, the fact that these pulse-like motions are typically associated with the regions of greatest damage made this issue particularly important.

Initially, our use of PGA and PGV for estimating intensities was also simply practical. We were only retrieving peak values from a large subset of the network, so it was impractical to compute more specific ground-motion parameters, such as average response spectral values, kinetic energy, cumulative absolute velocities (CAV, EPRI, 1991), or the JMA intensity algorithm (JMA, 1996) for example. However, because near-source strong ground-motions are often dominated by short-duration, pulse-like ground-motions (usually associated with source directivity), PGV does appear to be a robust measure of intensity for strong shaking. In other words, the kinetic energy (proportional to velocity squared) available for damage is well characterized by PGV. In addition, the close correspondence of the JMA intensities and peak ground velocity (Kaezashi and Kaneko, 1997) indicates that our use of peak ground velocities for higher intensities is consistent with the algorithm used by JMA. More recent work by Wu and others (2003) indicates a very good correspondence of PGV and damage for data collected on the island of Taiwan, which included high-quality loss data and densely sampled strong motion observations for the 1999 Chi-Chi earthquake. Finally, consideration in the choice of peak ground-motion values, rather than derived parameters, is the ease of relating intensity directly to simple ground-motion observables.

Nonetheless, for large distant earthquakes, the peak values will be less informative, and duration and spectral content may become key parameters. Although we may eventually adopt corrections for these situations, it is difficult to assign intensities in such cases. For instance, what is the intensity in the zone of Mexico City where numerous high-rises collapsed during the 1985 Michoacan earthquake? It was obviously high intensity shaking for high-rise buildings. However, the majority of smaller buildings were unaffected, indicating much lower intensity. Whereas the peak ground velocities were moderate and would imply I_{mm} VIII, resonance and duration conspired to cause a more substantial disaster.   Although this is, in part, a shortcoming of using peak parameters alone, it is more a limitation imposed by simplifying the complexity of ground-motions into a single parameter. Therefore, in addition to providing peak ground-motion values and intensity, we are also producing spectral response maps (for 0.3, 1.0, and 3.0 s). Users who can take advantage of this information for loss estimation will have a clearer picture than can be provided with maps of PGA and PGV alone. However, as discussed earlier, a simple intensity map is extremely useful for the overwhelming majority of users, which includes the general public and many involved with the initial emergency response.

We have also not yet addressed the potential for severe site effects and liquefaction of soft soil in California (NEHRP categories DE and E) such as in the Los Angeles Harbor region, much of the San Francisco Bay area, and along former and current river channels. Additional and significant losses can also result from down-slope ground deformation. For example, much of the losses in the greater Anchorage area during the 1964 Alaskan earthquake resulted from such movement and not from direct shaking damage.   Estimated intensities derived from peak velocity will not be sufficient for recognizing such effects and the increased effective intensity due directly to ground failure.

 Not only are we limited by the lack of sufficiently detailed geologic maps of such areas, but also the connection between the surface geology, the site amplification, and ground failure is not fully established for strong motions. Similarly, basin edge effects are not included, and differences between very deep basin and shallow basin sites are not yet distinguished. In addition, only peak values have been considered here; site resonance is not yet considered. Shaking duration has also not yet been included, though it may be important under certain circumstances. For instance, currently, we may underestimate the extent of damage (in terms of instrumental intensity) in Los Angeles for a great San Andreas event because only peak amplitude is considered. Similarly, intensities may be underestimated in Anchorage for a repeat of the great 1964 (magnitude 9.2) Alaska earthquake basing them on peak amplitude alone and not considering effects of long duration (particularly on ground failure), but currently there is little empirical constraint upon which to base a modification to the instrumental intensity computation for such an event. For such an earthquake, evaluation of the response spectral map may give more reliable estimates of potential damage.

The peak ground-motion versus intensity correlation is based on observations collected from recent California earthquakes. Hence, this relationship is subject to revision for other ANSS regions and to accommodate additional observations. At present, there is little data to correlate lower intensity values and recorded ground-motions because most of the ground-motion data are for larger earthquakes, and intensity data are not typically collected for smaller events, until recently. In addition, the calibration we have is primarily for analog recordings, so the noise level is high, especially for low amplitude (once-integrated) velocity seismograms. The digital data now being collected within ANSS regions will be more useful in calibrating against intensity at lower amplitudes. We are also collecting intensity measurements at near-station locations through voluntary response on the Internet (Wald and others, 1999c; URL http://pasadena.wr.usgs.gov/ shake). The combination of assigning intensities for low shaking levels with digital recordings will help constrain the relationship between acceleration, velocity, and intensity at the lowest values.

Naturally though, we are most concerned about accurately portraying the highest intensities. For example, approximately 86 percent of the residential losses in the Northridge earthquake occurred in the intensity VII-IX region (Kircher and others, 1997, p. 714). Intensity IX was the largest mapped value for that event. Interestingly, though, whereas the main emphasis of ShakeMap is to provide information about shaking for damaging earthquakes where the pattern of shaking can be quite complex, there has been widespread interest in viewing maps for smaller earthquakes, which are, nonetheless, widely felt. We generate ShakeMap for all earthquakes in California above magnitude 3.5-4.0, because the felt area for the smaller events is usually nominal. However, for several notable earthquakes in the magnitude 3.0 to 3.5 range, there has been a substantial demand for rapid display of the shaking pattern and so we have provided maps for these events as well. The advantage in providing ShakeMap for non-damaging earthquakes is twofold. First, we gain experience processing, calibrating, and checking our system by responding to small events daily to weekly, rather than on the very infrequent basis allowed by the occurrence of moderate to large earthquakes. Second, the user groups (which include emergency response agencies, utilities, the media, scientists, and the general public) are afforded the opportunity to become familiar with the maps and to test their response on a more regular basis.

2.6.2 Adding New Parameters

We are constantly re-evaluating or considering the use of additional ground-motion parameters, or intensity measures, for ShakeMap. However, any such additions cannot be made lightly. In part, this is due to the fact that the seismic network processing streams that produce parametric data for ShakeMap in different ANSS regions vary significantly. Indeed, even within the southern California region, ShakeMap data is produced both in real time with recursive filtering as well as with rapid post-processing and this is done by three different agencies. Mandating changes in such systems is not straightforward. Likewise, the addition of parameters in the processing stream not only takes more processing time, but we also like to limit the number of maps due to computational, bookkeeping, and storage efficiency considerations.

Candidates for additional parameters include energy or comparable measures (like cumulative average velocity, CAV) that include effects of duration and vector-based measures (e.g., Safak, 2000). However, ongoing engineering and loss-estimation research has not led to a obvious candidate that would justify overcoming the aforementioned obstacles so they have not warranted serious consideration at this time.

2.7 ShakeMap Uncertainty
[TBS]

2.7.1 Factors Contributing to Uncertainty
[TBS]

2.7.2 Quantifying Uncertainty
[TBS]

2.7.3 Examples for Significant and Scenario Earthquakes
[TBS]

2.8 Recent Example ShakeMaps

In this section we highlight ShakeMaps made for significant earthquakes in the past several years. These and other examples are best viewed interactively online on the ShakeMap Web pages (http://earthquake.usgs.gov/shakemap). Links found on the ShakeMap Web pages contain an archive of all ShakeMaps made to date as well as for major events that occurred prior to the advent of the current digital seismic networks and ShakeMap. These earlier events, e.g., the 1994 Northridge earthquake, were produced with the existing analog data recorded at the time, which were processed using the current ShakeMap tools and methodology.

2.8.1 1999 Hector Mine, California Earthquake

ShakeMaps have been generated in southern California because March 1997. The largest event to be recorded by the new TriNet system and mapped using ShakeMap was the October 16, 1999, magnitude 7.1 Hector Mine earthquake (Figure 2.7). Fortunately, the earthquake occurred in a remote area of the Mojave Desert, so little damage and few injuries were reported. Nevertheless, it was a good opportunity to evaluate the network and test the timeliness and quality of its products. Because the event occurred in a sparsely populated region, the spacing of seismic stations in the near-fault region was also sparse.

The performance of ShakeMap could be assessed under conditions that might prevail in a more urban earthquake for which near-fault stations might not immediately report due to power or communications failures. The TriNet real-time system determined a magnitude (energy magnitude) of 7.0 within 1 minute of the event, and ShakeMap was successfully produced and distributed within 4 minutes. The ground-motion from the Hector Mine event was widely felt in urban Los Angeles and, based on past experience, responders, the media and public had legitimate concerns regarding its source and potential damage. The ShakeMap provided rapid evidence that large-scale emergency response mobilization was unnecessary. The ShakeMap also highlighted areas of amplified ground-motion in the Coachella Valley and focused attention on numerous triggered events under the Salton Sea that were within 2 km of the San Andreas fault.

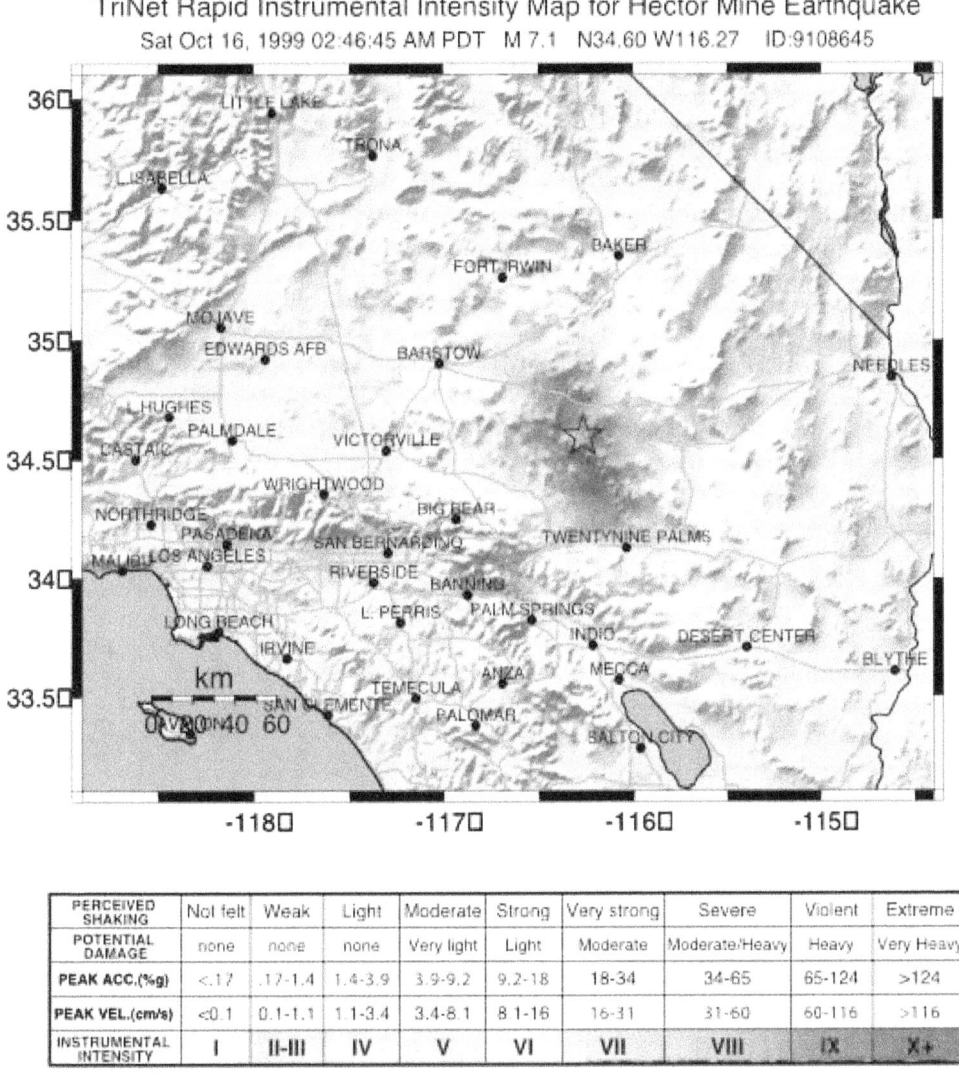

Figure 2.7   Instrumental Intensity ShakeMap for the October 16, 1999 magnitude 7.1 Hector Mine, California Earthquake.

2.8.2 2000 Napa Valley (Yountville), California Earthquake

Although moderate in size at magnitude 5.1, the September 3, 2000 Yountville earthquake caused significant damage in the city of Napa. The event occurred in the mountains 6 miles northwest of the city of Napa, near Yountville, California. As shown in Figure 2.8, the strongest shaking recorded was just north of the city of Napa. The recorded acceleration there was 50 percent of the force of gravity, rather high for this magnitude, but consistent with the significant damage that the city suffered.

Although earthquake shaking levels depend predominantly on the distance from the earthquake source, the high level of ground shaking in Napa appears to have been controlled by two other factors: first, the amplification of shaking by young sediments along the Napa River which shows as a topographic low on the ShakeMap intensity Map (Figure 2.8) and second, the focusing of strong motion to the southeast, the direction the earthquake rupture appears to have propagated. The offset of the strongest shaking to the southeast from the epicenter, and the amplification within the basin of sediments underlying Napa and along the northern shore of San Pablo Bay are also clear on the map of instrumental intensity.

ShakeMap quality strong motion instrumentation coverage in the San Francisco Bay area has also substantially improved because the 2000 Napa earthquake, so future earthquakes will have substantially better station control.

USGS/UCB/CDMG Rapid Instrumental Intensity Map for Yountville Earthquake
Sun Sep 3, 2000 01:36:30 AM PDT   M 5.1   N38.38 W122.41   ID:51101203

PERCEIVED SHAKING	Not felt	Weak	Light	Moderate	Strong	Very strong	Severe	Violent	Extreme
POTENTIAL DAMAGE	none	none	none	Very light	Light	Moderate	Moderate/Heavy	Heavy	Very Heavy
PEAK ACC.(%g)	<.17	.17-1.4	1.4-3.9	3.9-9.2	9.2-18	18-34	34-65	65-124	>124
PEAK VEL.(cm/s)	<0.1	0.1-1.1	1.1-3.4	3.4-8.1	8.1-16	16-31	31-60	60-116	>116
INSTRUMENTAL INTENSITY	I	II-III	IV	V	VI	VII	VIII	IX	X+

Figure 2.8   Instrumental Intensity ShakeMap for the magnitude 5.1 Napa Valley ("Yountville") earthquake on September 3, 2000.

2.8.3 2001 Seattle (Nisqually), Washington Earthquake

Figure 2.9 shows an example of a ShakeMap for one of the largest events to date to occur in a region of the country outside of California.   Although the 2001 Nisqually, Washington earthquake was of comparable magnitude to the 1994 Northridge earthquake, the depth of the rupture was much greater—near 50 km.   In contrast, the Northridge earthquake rupture was as shallow as 5 km.   Primarily as a result of this greater depth, the Nisqually earthquake caused approximately $0.3 billion of damage compared to $40 billion in losses due to the Northridge earthquake.

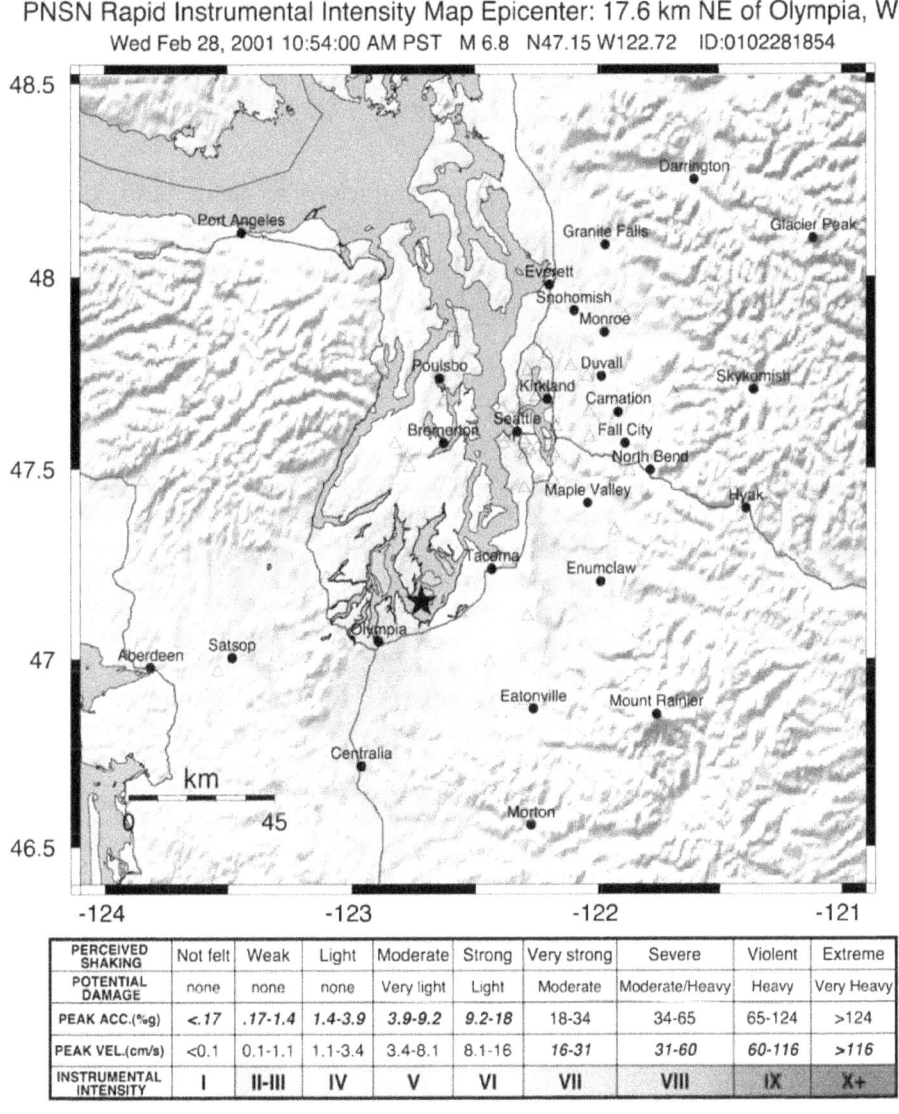

PERCEIVED SHAKING	Not felt	Weak	Light	Moderate	Strong	Very strong	Severe	Violent	Extreme
POTENTIAL DAMAGE	none	none	none	Very light	Light	Moderate	Moderate/Heavy	Heavy	Very Heavy
PEAK ACC.(%g)	<.17	.17-1.4	1.4-3.9	3.9-9.2	9.2-18	18-34	34-65	65-124	>124
PEAK VEL.(cm/s)	<0.1	0.1-1.1	1.1-3.4	3.4-8.1	8.1-16	16-31	31-60	60-116	>116
INSTRUMENTAL INTENSITY	I	II-III	IV	V	VI	VII	VIII	IX	X+

Figure 2.9   Example ShakeMap in the Pacific Northwest ANSS Region for the 2001 Nisqually, Washington (M6.8) earthquake. Open triangles depict station locations. Note correspondence of intensity of shaking and basin and lowland areas as revealed by the topographic base map.

The Nisqually earthquake occurred shortly after a major upgrade to the seismic network in the ANSS Pacific Northwest region, and the ShakeMap system in the Seattle region was installed but not fully operable at the time of the quake. Nonetheless, with substantial late-night efforts, ShakeMaps were made available within a day of the event. The ShakeMap in Figure 2.9 highlights the utility of comparing shaking intensity atop topographic relief. Because the topography serves as a proxy for site conditions (basins are typically flat, low-lying areas and steep mountains typically are rock), areas of amplified shaking usually correlate well with areas of low relief.

2.9 Regional ShakeMap Specifications

In this section we summarize specific customization employed for ShakeMap systems running or in development throughout the ANSS regions nationwide. Although we developed ShakeMap with portability in mind, region-specific issues need to be addressed as a part of the installation. To add a new region, the following criteria must be met:

1) *Parametric Data.* Peak ground-motions for both horizontal components of motion must be rapidly available following significant earthquakes. PGA and PGV are required (instrumental intensity is derived from these) and response spectral accelerations at 0.3,1.0, and 3.0 s are highly recommended. These parametric data can be unassociated as long as individual station files contain timing information, but preferably they are consolidated into a flat file (later converted to XML format) or, most preferable, loaded directly into a relational database for query from ShakeMap software upon being alarmed for an event.

2) *Mapping Files for Coverage Area.* The region over which ShakeMap can be properly constrained must be ascertained, and GMT formatted map files (roads, topography, cities, etc.) need to be collected for this region.

3) *Geology and Site Corrections.* ShakeMap requires a uniformly spaced grid of site conditions over the coverage area from which to make site corrections when performing interpolations between stations. We rely on NEHRP Classification (A-E, given as an associated average 30m shear velocity) and their corresponding amplification factors. Typically, site conditions are derived from a GIS-based geology map (or at least digital) that can be correlated appropriately with NEHRP site classifications.

4) *Distance-Attenuation Relations.* Ground-motion attenuation relationships (used for infilling data gaps) must be suitable for the regional attenuation and potential earthquake source locations and types. For example, for the Pacific Northwest, appropriate crustal

and subduction event equations are required. New relations can be easily added as PERL modules.

2.9.1 California

Efforts are underway to integrate the northern and southern California networks into the California Integrated Seismic Network (CISN). Under CISN plans, ShakeMap will be made more robust through remote, backup generation at northern and southern California operations centers. CISN will be a single region representing California in the ANSS, and effort to further integrate seismic monitoring throughout the entire United States.

2.9.1.1 Southern California

Coverage Area. Southern California ShakeMap is generated in the same region defined by the traditional authoritative earthquake-reporting region of southern California. Seismically, California is divided into northern and southern by the "Gutenberg-Byerly" line, an historic imaginary straight line agreed upon by Caltech and Berkeley in the early days of reporting earthquakes.

Triggering and Data Flow. ShakeMap triggering is in the form of an alarm message from USGS-Caltech real time network. An alarm is issued to ShakeMap once parametric data is available in the southern California Earthquake Data Center (SCEDC) Oracle database. Data flow in southern California is addressed in section 1.3.1 and the station distribution is shown in Figure 2.1.

Site Condition Map. The site condition map for southern California is shown in Figure 2.3 and is addressed in section 1.4.3.

Attenuation Relationships: Joyner and others (1997) is used for events larger than magnitude 5.5. For events of magnitude 5.0 and smaller, we use the equations derived specifically for southern California from a compilation of events with magnitudes ranging from 3.5. to 5.0. See Appendix A for more details.

Other Local Characteristics:
[TBS]

2.9.1.2 Northern California

Coverage Area. Northern California ShakeMap is generated in the same region defined by the traditional authoritative earthquake-reporting region of northern California (shown in Figure 2.1).

Triggering and Data Flow.
[TBS]

Site Condition Map. The site condition map for southern California is shown in Figure 2.3 and is addressed in section 1.4.2. For the San Francisco Bay area, however, the more detailed map of [Wentworth and others, 199?] is used and replaces the statewide map of Wills and others (2000).

Attenuation Relationships: Joyner and others (1997) is used for events larger than 5.5. For events 5.0 and smaller, Boatwright and others (2003) derived equations specifically for northern California from a compilation of events with magnitudes ranging from 3.5. to 5.0.

Other Local Characteristics: Backup in northern California is done with duplicate systems running in Menlo Park and at U.C. Berkeley.

2.9.2 Pacific Northwest

Coverage Area. [TBS]

Triggering and Data Flow. [TBS]

Site Condition Map. [TBS]

Attenuation Relationships. Joyner and others (1997) is used for crustal (shallow) earthquakes. For deeper events, Youngs et al, (1997) is employed with coefficients for intraslab and interplate events assigned by choosing default event depth ranges. The defaults can also be manually overridden once independent information about the source is known. See Appendix A for more details.

Other Local Characteristics: [TBS]

2.9.3 Intermountain West

2.9.3.1 Utah

Coverage Area. The University of Utah currently generates automatic ShakeMaps for earthquakes occurring in the Wasatch Front urban corridor in northern Utah (Figure 2.10). Different magnitude thresholds reflect differences in station coverage. The majority of the urban strong-motion stations are located in the Wasatch Front urban corridor (red box), where approximately 80 percent of the state's population lives astride the Wasatch fault.

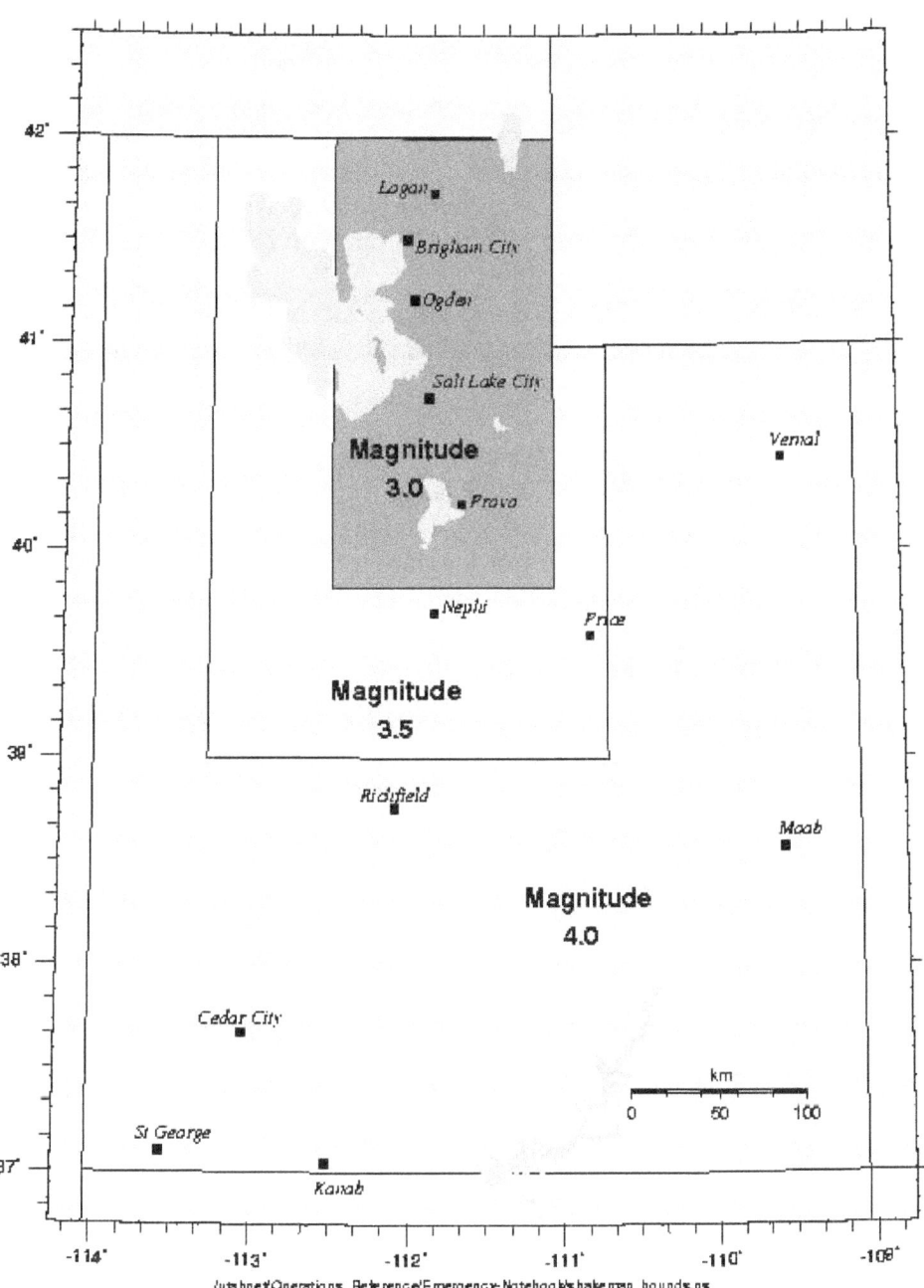

SHAKEMAP MAGNITUDE THRESHOLDS

Figure 2.10 Region and minimum magnitude thresholds for producing ShakeMaps in Utah. Earthquakes with magnitudes larger than 5.0 outside of the Utah region and within 120 km of a Utah seismic station will also generate ShakeMaps. However, the epicenter will not appear on the map, only the resulting ground-motion.

Triggering and Data Flow. Using the Earthworm software package (see http://folkworm.ceri.memphis.edu/ew-doc/) the University of Utah Seismograph Stations

(UUSS) collects data in near-real-time from seismic stations throughout the state and surrounding regions. Using this data, Earthworm associates seismic events recorded at different stations and calculates a location and magnitude. For earthquakes above magnitude 2.96, Earthworm generates a ShakeMap compatible XML formatted file containing parametric peak ground acceleration (PGA), peak ground velocity (PGV), and 5 percent-damped pseudo-acceleration (PSA) values from the horizontal components from up to 96 strong-motion and broadband instruments (Figure 2.11). Earthworm also writes the earthquake source information to an XML file. These files are placed in a directory that ShakeMap monitors. Once the two files for an event appear in the directory, a queuing program is run to determine if ShakeMap should start. The queuing program also prioritizes events by size and distance to the population centers. This is particularly useful in the case of aftershocks or swarms. Additional data from up to 10 stations maintained by the USGS National Strong Motion Program are manually merged into the XML file as data become available.

Once the two files for an event appear in the directory, a queuing program is run to determine if a ShakeMap should start. Depending on the distance to the major population centers different magnitude thresholds are used for actually producing maps (Figure 2.10). For instance, ShakeMaps are produced for earthquakes of magnitude 3.0 or larger occurring in the densely populated region from Logan to Nephi. Outside of that region the minimum magnitude is 3.5. In addition, the queuing program is configured to prioritize events by size and distance to the population centers. This is particularly useful in the case of aftershocks or swarms.

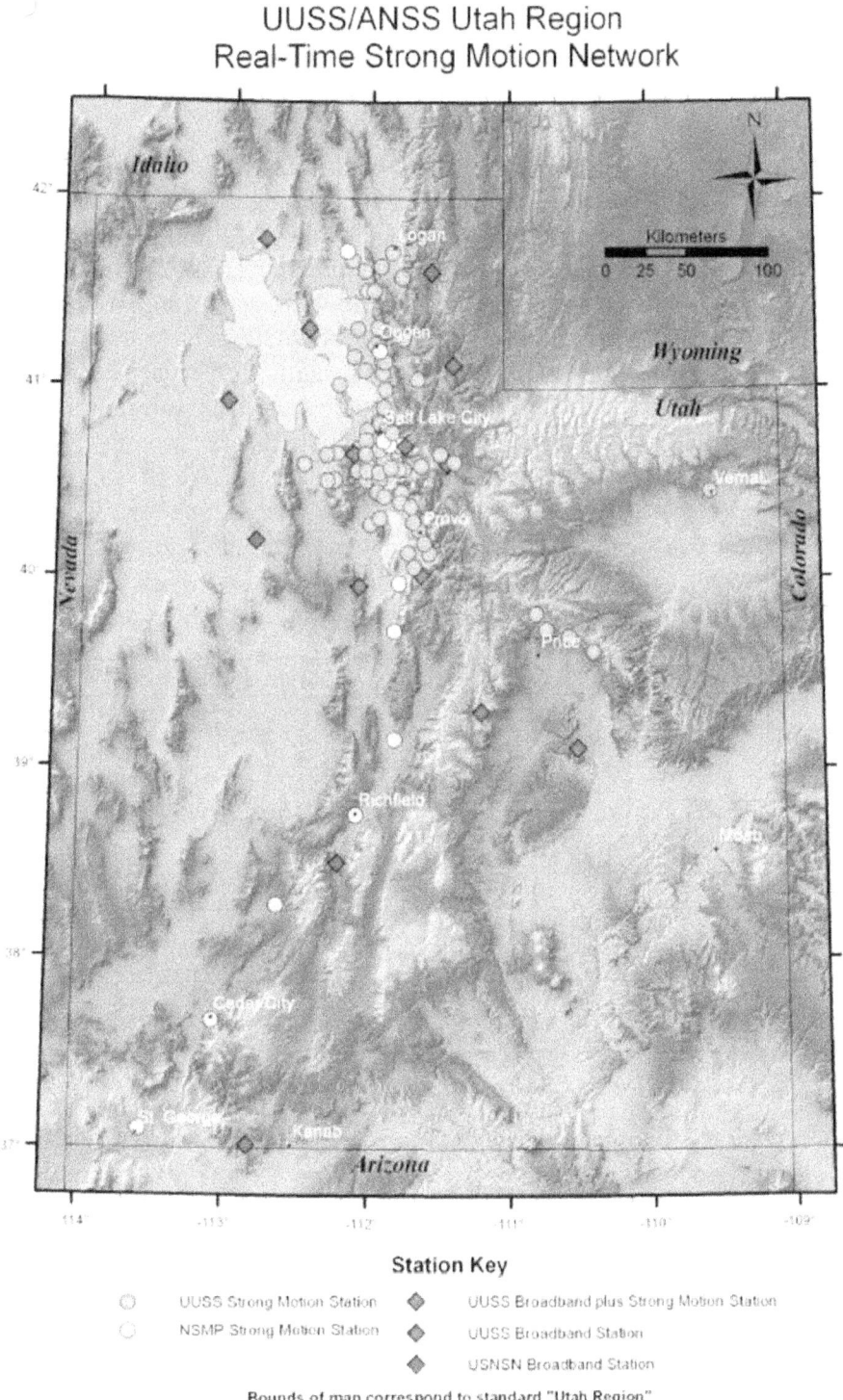

Figure 2.11 The Advanced National Seismograph ShakeMap network for the Wasatch Front Urban Corridor, Utah as of September 30, 2005.

Site Condition Map. Once the ground motion is calculated for "rock," we apply site amplification factors to correct for the local geology. These factors were calculated using equations 7a and 7b from the Appendix in Borcherdt (1994) and a reference velocity of 910 m/sec. The average shear velocity in the upper 30 meters (Vs30) for local geologic units and corresponding amplification factors are in Table 2.3. Detailed geologic mapping and grouping by Vs30 for the Utah ShakeMap region was done by the Utah Geological Survey (Ashland, 2001; Ashland and McDonald, 2003; G. N. MacDonald written communication, 2005). The mapping was done at two scales: 1:500,000 for the state and 1:250,000 for the region from Provo to Brigham City. In the finely mapped region, the grouping of Vs30 units consists of 4 distinct quaternary soil units—Q01, Q02, Q03, Q05, and 3 rock units -- Tertiary, Mesozoic, and Paleozoic rock units. In the larger scale regions an average Quaternary soil unit and the three rock units were used (Figure 2.12). Although this is the mapping that is currently available, one area of concern is that all of the Vs30 measurements were made in Lake Bonneville deposits. Mapping Vs30 values from Lake Bonneville deposits to more general quaternary deposits may not be appropriate. Refining the Vs30 measurements and site amplification factors are active areas of research in the region.

Class	Vs30		Short-Period (PGA)				Mid-Period (PGV)			
			150	250	350		150	250	350
P	2197		0.73	0.80	0.92	1.05		0.56	0.59	0.63	0.67
M	1449		0.85	0.89	0.95	1.02		0.74	0.76	0.78	0.81
T	1023		0.96	0.97	0.99	1.01		0.93	0.93	0.94	0.95
Q	234		1.61	1.40	1.15	0.93		2.42	2.26	2.05	1.84
Q01	199		1.70	1.46	1.16	0.93		2.69	2.49	2.24	1.98
Q02	301		1.47	1.32	1.12	0.95		2.05	1.94	1.80	1.65
Q03	387		1.35	1.24	1.09	0.96		1.74	1.67	1.57	1.47
Q04	437		1.29	1.20	1.08	0.96		1.61	1.55	1.48	1.39
Q05	486		1.25	1.17	1.06	0.97		1.50	1.46	1.39	1.33

Table 2.3 Site Correction Amplification factors. Short-Period (.1 to .5 sec) factors from equation 7a, Mid-Period (.4 to 2. sec) from equation 7b of Borcherdt (1994). Class is geologic grouping done by Ashland (2001); Vs30 is the average shear-wave velocity in the upper 30 m (m/s) and PGA is cutoff input PGA in gals.

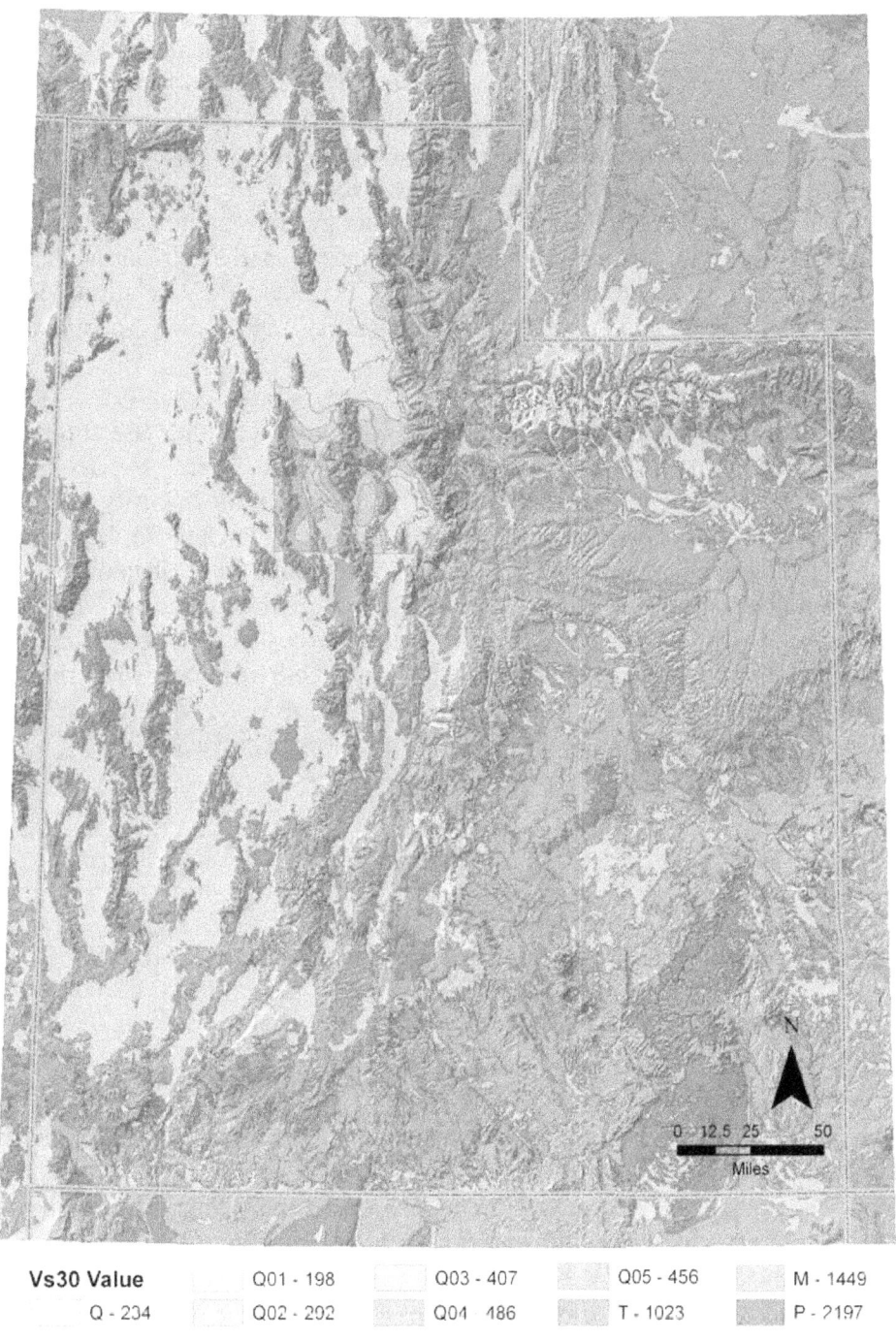

Vs30 Value	Q01 - 198	Q03 - 407	Q05 - 456	M - 1449
Q - 234	Q02 - 292	Q04 486	T - 1023	P - 2197

Figure 2.12 Wasatch Front Site Condition Map based on geology and Vs30. Adapted from Ashland (2001) and Ashland and McDonald (2003). The colors correspond to Vs30 groupings. Geologic mapping was done at two scales: Wasatch Front 1:250,000, rest of the region 1:500,000.

Attenuation Relationships. To approximate the ground motion to "rock" in regions of sparse data coverage, we use attenuation relations from Pankow and Pechmann (2004) to calculate the ground motion to a reference rock site. The PGA and PSV relations for rock in Pankow and Pechmann (2004) are similar to those reported in Spudich *et al.* (1999) except that the reported bias at rock sites has been corrected. The PGV relation in Pankow and Pechmann (2004) was developed using PGV data collected for the same events as in Spudich *et al.* (1999; Paul Spudich, personal communication). All of these relations are appropriate for extensional tectonic regimes, for earthquakes with magnitudes between 5.0 and 7.7, and event-station distances < 100 km. For earthquakes with magnitudes < 5.0, we use PGA and PGV relations developed for Southern California (V. Quitoriano, written communication, 2002). See Appendix A for more details.

Other Local Characteristics. Once the ShakeMaps are produced they are transferred to the UUSS web page (http://www.quake.seis.edu) and the USGS web page (http://www.earthquake.usgs.gov). In addition, a JPEG version of the intensity map is emailed to Utah Division of Emergency Services and Homeland Security, the Utah Geological Survey, and duty seismologists' home email accounts. Generally, ShakeMaps are reviewed for quality within the first few hours of posting. Within several days of the earthquake, the data are manually reprocessed and reviewed. At this point the map will be re-posted and the disclaimer flag "Not reviewed by human" is removed. It is worth noting, UUSS runs two duplicate systems of Earthworm and ShakeMap. They are configured so that in case of system failure on the active machine the backup can be smoothly transitioned without loss of service.

2.9.3.2 Nevada

[TBS]

Status. Currently enhancing station distribution and testing ShakeMap software.

2.9.4 Mid-America

Coverage Area. The Center for Earthquake Research and Information (CERI), University of Memphis, will generate automatic ShakeMaps for earthquakes occurring in the New Madrid Seismic Zone. The trigger area is located in the Upper Mississippi Embayment of the central United States and is centered on the New Madrid seismic zone (Figure 2.13). It covers a four by four degree area from 92°W to 88°W and 35°N to 39°N and is approximately 450 km by 450 km or 202,500 square kilometers. The area encompasses 6 states and the major metropolitan areas of Memphis, Tennessee and Saint Louis, Missouri.

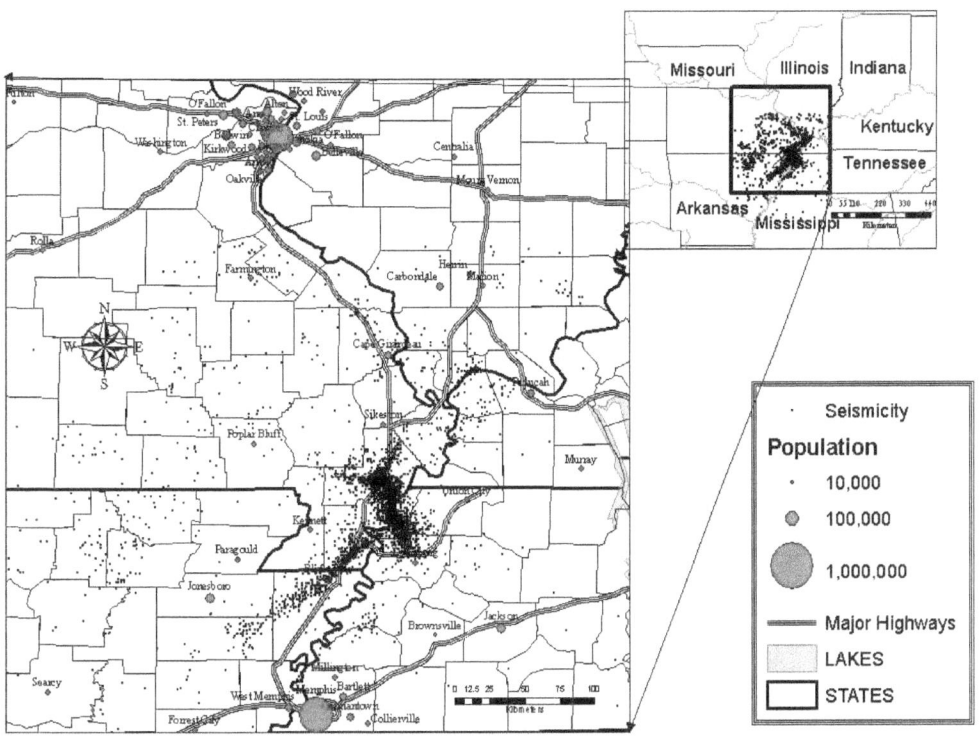

Figure 2.13   The map outline is the regional extent for the production of ShakeMap maps. Earthquakes located within this region, with magnitudes larger than 3.0, generate automatic ShakeMaps. The New Madrid Seismic Zone is defined by the seismicity denoted here as black dots.

Triggering and Data Flow.   Using the Earthworm software package (see http://folkworm.ceri.memphis.edu/ew-doc) CERI collects data in real time from seismic stations throughout the surrounding region. Using this data, Earthworm associates seismic events recorded at different stations and calculates a location and magnitude. For earthquakes above magnitude 3.0, Earthworm also calculates parametric peak ground acceleration (PGA), peak ground velocity (PGV), and 5 percent-damped pseudo-acceleration (PSA) values from the horizontal components from up to 56 strong-motion and broadband instruments (Figure 2.14). This information is written to a ShakeMap compatible XML formatted file. These files are automatically placed in a directory that ShakeMap monitors.

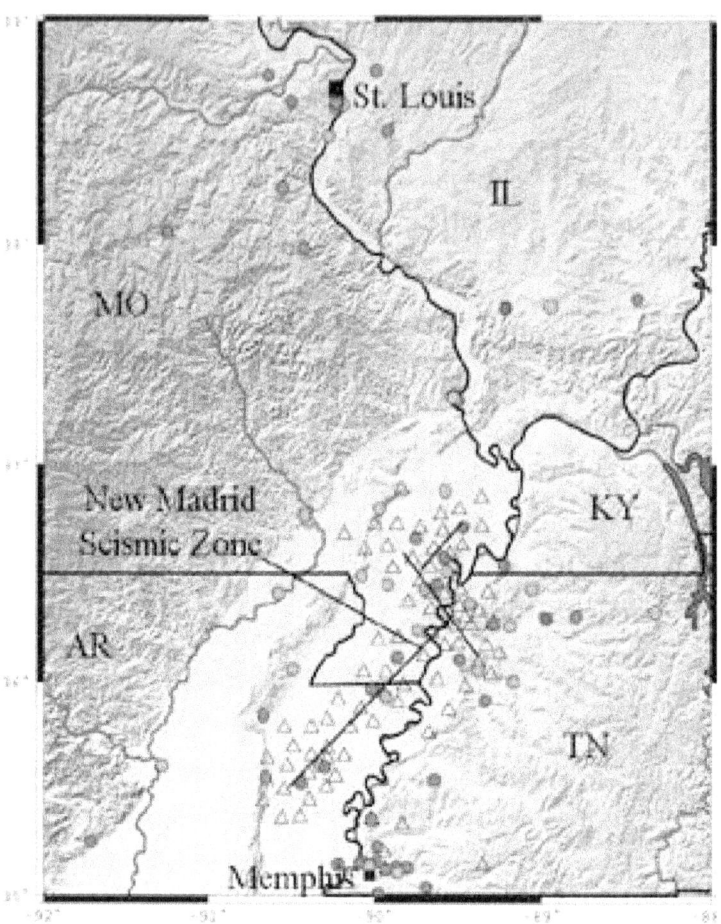

Figure 2.14   The New Madrid Cooperative Seismic network for the Upper Mississippi
Embayment, Mid-America, as of July 2005. University of Memphis, CERI and St. Louis
University broadband and strong motion stations are in red, short period seismometers in
open triangles, the U.S. National Seismic Network (USNSN) in dark blue, the National
Strong Motion Program (NSMP) in green. Stations operated by CERI, SLU, and USNSN,
are recorded at CERI in real-time. Short period stations are used for location purposes
only.

Once the two files for an event appear in the directory, a ShakeMap queuing program is run to
determine if a ShakeMap should start. A local magnitude threshold of 3.0 is used for producing
maps (Figure 2.13). In addition, the queuing program is configured to prioritize events by size
and distance to the population centers. This is particularly useful in the case of aftershocks or
swarms.

Site Condition Map.   The ground-motion is calculated for "rock," and a site amplification factor
is applied to correct for the effects of the local geology. These factors were calculated using
equations 7a and 7b from the Appendix in Borcherdt (1994) and a reference velocity of 750 m/s.
The National Earthquake Hazard Reduction Program's (NEHRP) system of soil classification
(FEMA, 1994) is the standard soil classification scheme used by the Mid-America region. This
methodology assigned soil classification letters of A, B, C, D, E1, E2, F1, F2, F3 and F4 as

defined by the soil's geological description, shear wave velocity, potential to liquefy and other engineering parameters (Table 2.4) (FEMA, 1994).

Table 2.4

Soil profile type classification for seismic amplification (FEMA, 1994).

Soil Type	General Description	Avg. Shear Wave Velocity (feet/s)	Avg. Shear Wave Velocity (m/s)	Avg. Blow Counts	Avg. Shear Strength (lbs/sq.ft.)
A	Hard Rock	> 5,000	> 1,500		
B	Rock	2,500 - 5,000	760 - 1,500		
C	Hard and/or stiff/very stiff soils; most gravels	1,200 - 2,500	360 - 760	> 50	2,000
D	Sands, silts and/or stiff/very stiff clays, some gravels	600 - 1,200	180 - 360	15 - 50	1,000 - 2,000
E	Small to moderate thickness (10 to 50 feet) soft to medium stiff clay, Plasticity Index > 20, water content > 40 percent	< 600	< 180	< 15	< 1,000
E_2	Large thickness (50 to 120 feet) soft to medium stiff clay Plasticity Index > 20, water content > 40 percent	< 600	< 180	< 15	< 1000
F_1	Soils vulnerable to potential failure or collapse under seismic loading such as liquefiable soils, quick and highly sensitive clays, collapsible weakly cemented soils.	By definition the F classification requires that a site dependent evaluation of the engineering parameters be conducted, as they do not fall into any of the other soil classifications.			
F_2	Peats and/or highly organic clays greater than 10 feet thick				
F_3	Very high plasticity clays greater than 25 feet thick with Plasticity Index > 75				
F_4	Very thick soft/medium stiff clays greater than 120 feet thick				

The Central United States Earthquake Consortium (CUSEC) Association of State Geologists assembled information on earthquake hazards for the New Madrid Seismic Zone of the CUSEC region. They developed a standard method to create a soil amplification potential map, showing the potential seismic shaking hazard due to soil types (Bauer *et. al.,* 2001). The map, Compilation of Databases and Map Preparation for Regional and Local Seismic Zonation Studies in the CUSEC Region (CUSEC Map), covered portions of the states of Arkansas, Illinois, Indiana, Kentucky, Mississippi, Missouri, Ohio and Tennessee, including the 1 x 2 degree (scale 1:250,000 or 1 inch = 3.9 miles) Belleville, Rolla, Vincennes, Evansville, Dyersburg, St. Louis,

Poplar Bluff, Blytheville, and Memphis quadrangles (Bauer *et. al.,* 2001). Geologic maps of surficial materials were used in combination with field measured shear wave velocities to classify the soils, according to the NEHRP soil classification schema (see above), for the upper 15 to 30 meters and the results were distributed on compact disc (Bauer *et. al.,* 2001). The Geographical Information System (GIS) format of the maps was used in the creation of the regional ShakeMap amplification factors.

One topic of concern is the soil type designation of "F" on the map pertains to liquefiable soils; ShakeMap makes no distinction for this soil type. In order to work around this problem the "F" designation was assigned an "E" designation. However, it should be noted that recent geophysical surveys by Street *et. al.,* (2004) showed that a section of the embayment designated by the CUSEC map as type "F" (assumed herein to be "E"), exhibited velocities of soil type "D". Additionally, since individual State Geological Surveys conducted independent assessments of their respective states, there were data discrepancies from state to state (Bauer, personal communication). This was evident when changes in soil types at the Arkansas, Missouri border (Figure 2.15) were observed. The average shear velocity in the upper 30 meters (Vs30) for local geologic units and corresponding amplification factors are shown in Table 2.5.

Average shear wave velocity for local geological units									
Class	Vs30	Short-Period (PGA)				Mid-Period (PGV)			
			150	250	350		150	250	350
B	1130	1.00	1.00	1.00	1.00	1.00	1.00	1.00	1.00
BC	750	1.15	1.11	1.04	0.98	1.31	1.28	1.24	1.20
C	560	1.28	1.19	1.07	0.97	1.58	1.52	1.45	1.37
CD	360	1.49	1.33	1.12	0.94	2.10	1.99	1.83	1.67
D	270	1.65	1.43	1.15	0.93	2.54	2.36	2.14	1.90
DE	180	1.90	1.58	1.20	0.91	3.30	3.01	2.65	2.29
E	180	1.90	1.58	1.20	0.91	3.30	3.01	2.65	2.29

Table 2.5 Site Correction Amplification factors. Short-Period (.1 to .5 sec) factors from equation 7a, Mid-Period (.4 to 2. sec) from equation 7b of Borcherdt (1994). Class is geologic grouping done by Bauer (2001); Vs30 is the average shear-wave velocity in the upper 30 m (m/s) and PGA is cutoff input PGA in gals.

The coverage area of the CUSEC map constrained the area for ShakeMap to accurately display amplified shaking. Therefore, the aerial extent of the CUSEC map is an area for future improvements. Recent geophysical and engineering velocity data on soil locations beyond the current maps should be incorporated into a new map of larger coverage area. The area to the south of Memphis, Tennessee in northern Mississippi and southern Arkansas should be included, as the population in this area is expanding rapidly (Figure 2.15).

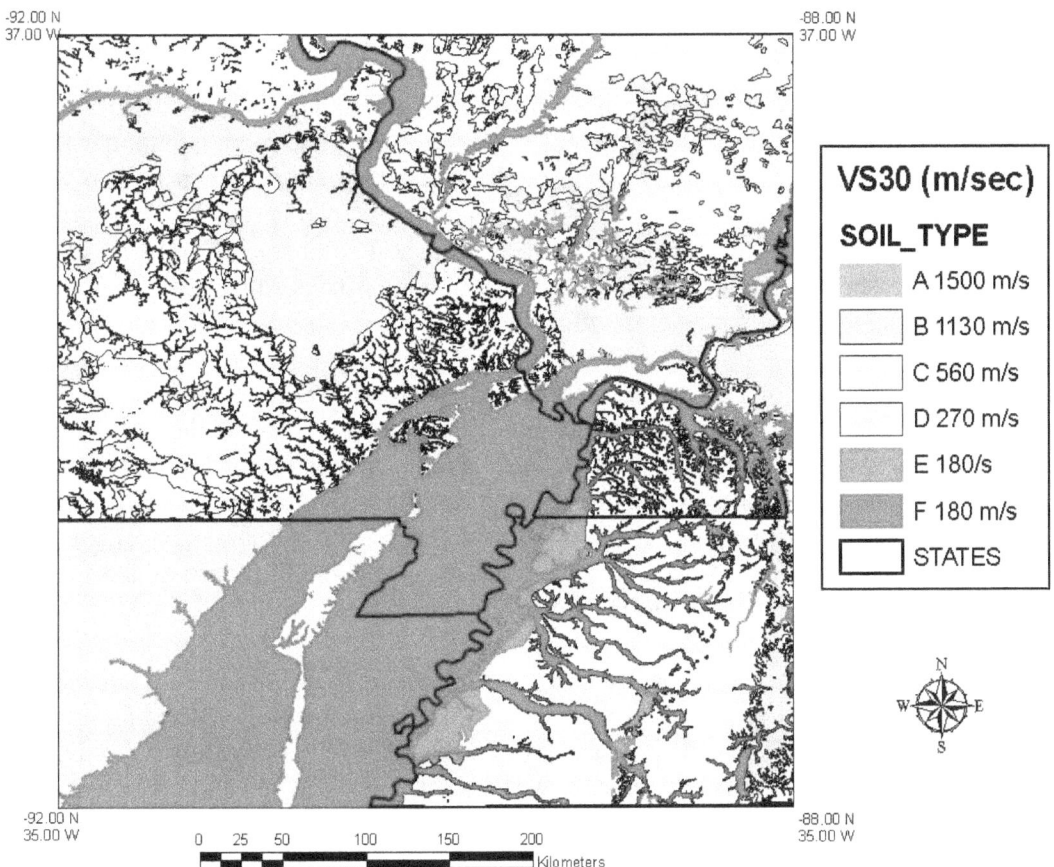

Figure 2.15   New Madrid Seismic Zone Site Condition Map based on geology and Vs30, from Bauer *et. al.* (2001). The colors correspond to Vs30 groupings. Final geologic mapping was done at 1:250,000.

Attenuation Relationships.   Earthquakes in the central and eastern United States are inherently different than those in the Western United States with regard to attenuation, energy release and characteristics of strong ground motion (e.g. McGuire, 1987). Therefore attenuation relationships calibrated for the Western United States will not adequately represent ground motions in the central and eastern United States (Kaka and Atkinson, 2004, Brackman, 2005).

Several researchers developed attenuation relationships for the Central United States (e.g. Boore and Atkinson, 1987; Toro and McGuire, 1987; Boore and Joyner, 1991; EPRI, 1993; Toro et al., 1997; Atkinson and Boore, 1997; Frankel et. al., 1996; Somerville et. al., 2001; Campbell, 2002; EPRI 2004: Kaka and Atkinson 2005). In order to implement a well-established, consensus-based attenuation relationship, the plan was to incorporate multiple weighted attenuation relations into ShakeMap in agreement with the CEUS Portion of Draft Versions of the 2002 Update of the National Seismic Hazards Maps (Frankel, 2002). The 2002 Hazard maps include the attenuation relations of Atkinson and Boore (1995), Toro et al. (1997), Frankel et al. (1996), Somerville et al. (2001) and Campbell (2002). However, until such time as software improvements are available, we instead use a single relationship that is most compatible with our needs and available data.

The majority of eastern United States attenuation relations are designed for magnitudes greater than six. Kaka and Atkinson (2005), in an attempt to model smaller and more common events, used data from central and eastern United States empirical databases in conjunction with modeled data from Atkinson and Boore (1995). The equation obtained is typically based on recorded ground motions of magnitudes less than five. Kaka and Atkinson, (2005) state that the relationship might under estimate peak ground motions for magnitudes equal to or greater than six, therefore, limiting the range to lower magnitudes.

The attenuation relationships of Toro et. al., (1997), Atkinson and Boore (1995) and Kaka and Atkinson, (2005) were tested for accuracy (Brackman, 2005). Results showed the attenuation relationship of Kaka and Atkinson (2005) to be in reasonable agreement with the Community Internet Intensity Maps with a minimal amount of over predicting (Brackman, 2005) for smaller events. The relationship of Toro et. al., (1997) was found sufficient for emergency response personnel to identify where the most intense damage has occurred and the approximate extent of damage (Brackman, 2005) for larger ground motions.

For the Upper Mississippi Embayment study area the relationship of Kaka and Atkinson (2005) should be used to predict peak ground motions for magnitudes at and below six and the relationship of Toro *et. al.,* (1997) should be used for earthquakes of magnitude greater than six. The relationships will need to be reassessed as new information is gathered and predictive models improve.

Instrumental intensity.   ShakeMap uses the Instrumental Intensity regression to map recorded and modeled peak ground motions to MMI. Wald *et al.,* (1999a) developed an instrumental intensity regression, for use specifically by ShakeMap locations in the Western United States. However, it has been recognized that intra-plate earthquakes, like those in the central and eastern United States, are associated with higher stresses and, in the near source these ground motions may be characterized by higher peak ground motions plus variable frequency content (Kanamori and Anderson, 1975). Atkinson (1993a) states that earthquakes recorded in California may have a lower frequency content than those recorded in the central and eastern United States, and therefore, PGV and PGA have a different meaning in the two regions. Kaka and Atkinson (2004) has been shown (Brackman, 2005) to be the best instrumental intensity regression for ShakeMap implementation in Mid America. Research to develop a relationship between PGV and MMI for the New Madrid region is ongoing (Atkinson, personal communication). A region specific regression would be a considerable advancement for ShakeMap, as it would give better constraints on MMI and peak ground motions. Since Kaka and Atkinson's (2004) regression for instrumental intensity has the ability to be corrected for magnitude and distance, additional programming should be done to incorporate this aspect into the existing software, increasing ShakeMap's accuracy.

Other Local Characteristics.   Automated generation of ShakeMap at CERI is in its infancy. After a reasonable period of testing and evaluation we will determine the most appropriate notification mechanisms and recipients.

2.9.5 Northeast

[TBS]

Status. Planning stages.

2.9.6 Alaska

[TBS]

Coverage Area. Fully operational but in test mode.

Triggering and Data Flow. Initial triggering will come from the Alaska Tsunami Warning Center (ATWC) via QDDS/QDM. Updates from either Alaska Earthquake Information Center (AEIC) or the National Earthquake Information Center (NEIC) will then take precedence depending on the authoritative region and network for the particular event.

Site Condition Map. [TBS]

Attenuation Relationships. Joyner and others (1997) is used for crustal (shallow) earthquakes. For deeper events, Youngs et al, (1997) is employed with coefficients for intraslab and interplate events assigned by choosing default event depth ranges. The defaults can also be manually overridden once independent information about the source is known. See Appendix A more details.

Other Local Characteristics: Run in Golden, CO at the USGS National Earthquake Information Center.

2.9.7 Hawaii

[TBS]

Status. Planning stages.

2.9.8 Puerto Rico and U.S. Territories

[TBS]

Status. Currently enhancing station distribution and testing ShakeMap software.

2.10 Scenario Earthquakes

In planning and coordinating emergency response, utilities, local government, and other organizations are best served by conducting training exercises based on realistic earthquake situations—ones that they are most likely to face. Scenario earthquakes can fill this role. The ShakeMap system can be used to map ground-motion estimates for earthquake scenarios as well as real data. Scenario maps can be used to examine exposure of structures, lifelines, utilities, and transportation conduits to specific potential earthquakes. ShakeMap Web pages now display selected earthquake scenarios and more events will be added as they are requested and produced.

ShakeMap earthquake scenarios are an integral part of emergency response planning in southern California where the ShakeMap system has been in place the longest. Primary users include city, county, state and federal government agencies (e.g., the California Office of Emergency Services, FEMA, the Army Corp of Engineers) and emergency response planners and managers for utilities, businesses, and other large organizations. Scenarios are particularly useful in planning and exercises when combined with loss estimation systems such as HAZUS and the Early Post-Earthquake Damage Assessment Tool (EPEDAT; Eguchi and others, 1997), which provide scenario-based estimates of social and economic impacts.

An unexpected, but very useful benefit of scenario generation is the added familiarity for those responsible for ShakeMap operations. Through the generation of many large events, a number of the ShakeMap configurations are adjusted and refined, allowing more rote response to real earthquakes. Again, this is one of the fundamental goals in creating scenarios: planning for and being ready for infrequent, but damaging earthquakes where timely and suitable response is mandated.

In this section we describe the procedures for generating and standardizing ShakeMap earthquake scenarios, with emphasis on differences with respect to real events for which maps are triggered automatically and constrained by strong motion observations. We also describe the technical and scientific rational for representing scenarios in the simplified form described below.

2.10.1   Generating Earthquake Scenarios

Given a selected event, we have developed tools to make it relatively easy to generate a ShakeMap earthquake scenario using the following steps: 1) Assume a particular fault or fault segment will (or did) rupture over a certain length and with a chosen magnitude, 2) Estimate the ground shaking at all locations over a chosen area surrounding the fault, and 3) Represent these motions visually by producing ShakeMaps and generating ground-motion input for loss estimation modeling (e.g., FEMA's HAZUS). At present, ground-motions are estimated using empirical attenuation relationships to estimate peak ground-motions on rock conditions. We then correct the amplitude at that location based on the local site soil (NEHRP) conditions as we do in the general ShakeMap interpolation scheme.   Finiteness is included explicitly, but directivity enters only through the empirical relations, though it too can be added explicitly as well. The choice of this representation is described below.

Our approach is simple and approximate. We account for fault finiteness by measuring the distance to the surface projection of the fault location (Joyner and Boore's distance definition), but in the default case we do not consider the direction of rupture nor do we modify the peak motions by a directivity term. Fault geometries are specified with a fault file that represents either the surface trace of the fault or the surface projection of the fault area. In either case the surface expression of the rupture is shown on the map as shown in Figure 2.16.

With this approach, the location of the earthquake epicenter does not have any effect on the resulting ground-motions; only the location and dimensions of the fault matter. If we were to add directivity to the calculations, than different choices of epicentral location would result in significantly different motions for the same magnitude earthquake and fault segment. Rather, our approach here is to show the average effect because it is difficult to justify a particular choice of hypocenter or to show the results for every possible hypocentral location. Our empirical predictive approach also only gives average peak ground-motion values so it does not account for all the expected variability in motions, other than the aforementioned site amplification variations. Actual ground-motions show significant variability for a given distance, magnitude, and site condition and, hence, the scenario ground-motions are more uniform than would be expected for a real earthquake. The true variations are partially attributable to 2D and 3D wave propagation, path effects (such as basin edge amplification and focusing), differences in motions among earthquakes of the same magnitude, and complex site effects are not accounted for with our methodology. For scenarios in which we wish to explore directivity explicitly, the Somerville (1997) regression is included in the ShakeMap package (see Appendix A).

As an example of the effectiveness of the scenario generation process, Figure 2.16 shows both the observed ShakeMap for the 1994 Northridge earthquake (left) and an estimated ShakeMap scenario (right) computed with the same earthquake source information assumed in the typical scenario calculations: the magnitude and geometry of the fault that slipped. In this case the dimensions of the Northridge rupture are known from analyses of the earthquake source (e.g., Wald and others, 1996).

In the current ShakeMap scenarios, we do not explicitly include the effects of rupture directivity, which has been shown to concentrate energy and the strongest shaking away from the hypocenter and in the direction that the fault rupture progresses. In Figure 2.16, the observed shaking from the Northridge earthquake (left) has more energy in the region northwest of the epicenter than the scenario version (right). This is due to the fact that the earthquake indeed exhibited northwestward directivity, and ShakeMap includes this only in an average sense in the predictions for the scenario. However, much of the shaking pattern is recovered just by knowing the dimensions of the fault that ruptured. In the case of strike slip earthquakes like the Newport-Inglewood and San Andreas fault (Ft. Tejon) scenarios, shown on the ShakeMap Scenario Web page archive, directivity can be quite severe, so depending on where the actual epicenter is, the shaking pattern might be skewed toward stronger shaking away from the epicenter than is shown in our scenarios.

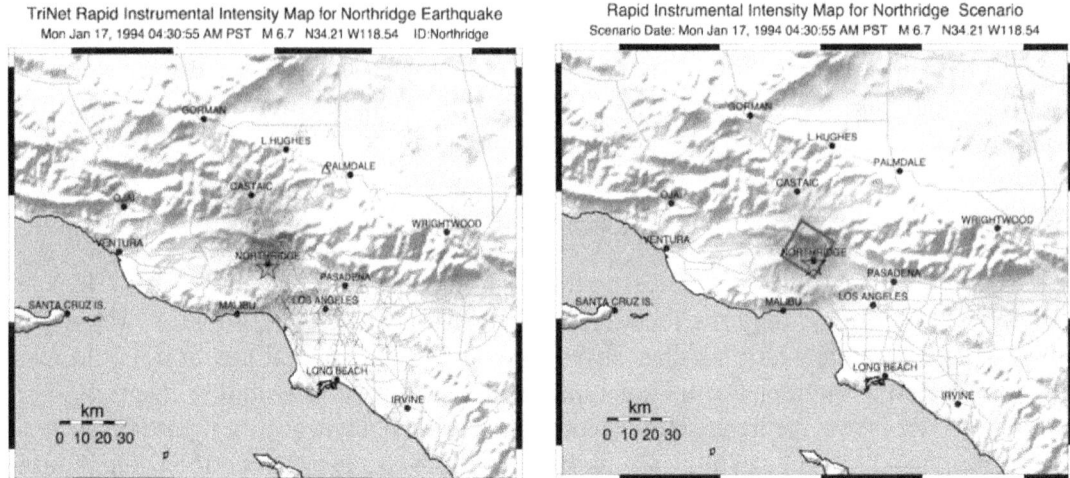

Figure 2.16 Northridge Earthquake ShakeMap (Left) and scenario earthquake (Right) for the Northridge earthquake made by assuming the correct magnitude and fault rupture area shown projected to the surface (black rectangle).

In terms of generating scenarios with the ShakeMap system, a number of specific considerations and a number of configuration changes are made for scenario events as opposed to actual events triggered by the network. For example, after generating a scenario for a major but hypothetical event, (obviously!) one does not want to automatically deliver the files to customers who are expecting real events. To avoid possible operator errors, all scenarios are tagged with the suffix "_se" in the event name. Such events are recognized by the processing software, which is configured to ignore steps normally taken for a real earthquake, unless manually overwritten.

Another obvious consideration for avoiding improper use of the scenario maps is noticeable and sufficiently redundant labeling of all Scenario maps (Figure 2.16).

2.10.2   Standardizing Earthquake Scenarios

The U.S. Geological Survey has evaluated the probabilistic hazard from active faults in the United States for the National Seismic Hazard Mapping Project. From these maps it is possible to prioritize the best scenario earthquakes to be used in planning exercises by considering the most likely candidate earthquake fault first, followed by the next likely, and so on. Such an analysis is easily accomplished by hazard deaggregation, in which the contributions of individual earthquakes to the total seismic hazard, their probability of occurrence and the severity of the ground-motions, are ranked. Using the individual components ("deaggregations") of these hazard maps, a user can properly select the appropriate scenarios given their location, regional extent, and specific planning requirements.

In California, the California Geological Survey (CGS) and the USGS have evaluated the probabilistic hazard from active faults in the state as part of the Probabilistic Seismic Hazard Assessment for the State of California described by Peterson and others (1996) and the National Seismic Hazard Mapping Project described by Frankel and others (1996). Currently, the

ShakeMap scenario events come directly out of the CGS catalog of fault source parameters that make up the statewide probabilistic seismic hazard assessment.

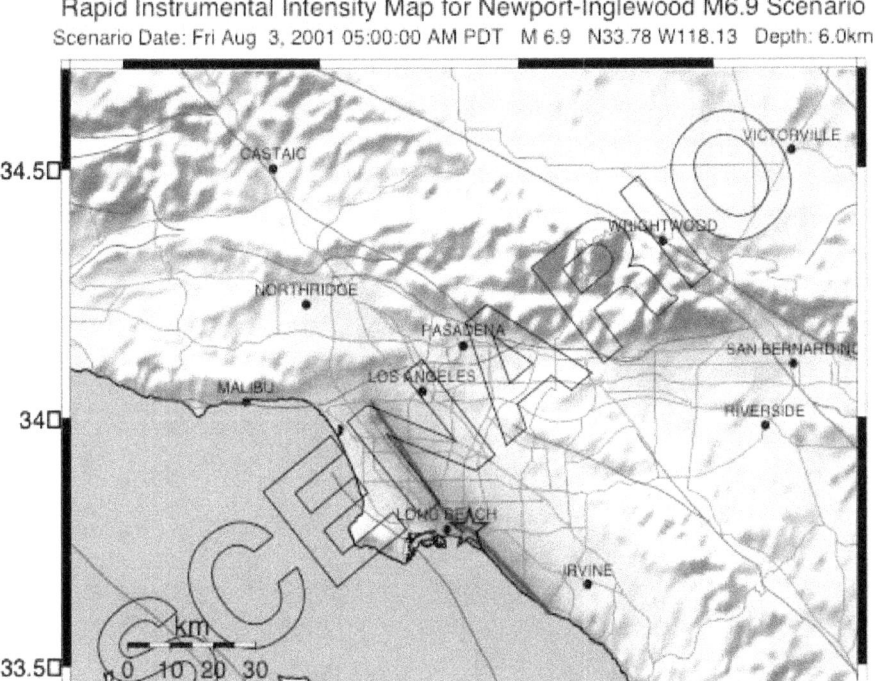

Figure 2.17 Example of a ShakeMap Scenario Earthquake for a hypothetical magnitude 6.9 earthquake on the Newport-Inglewood fault near Los Angeles. This scenario represents one the most destructive earthquakes that could impact the region. Note the redundant occurrences of the word "Scenario" to avoid confusion with an actual earthquake.

Scenarios are of fundamental interest to scientific audiences interested in the nature of the ground shaking likely experienced in past earthquakes as well as the possible effects due to rupture on known faults in the future. In addition, more detailed and careful analysis of the ground-motion time histories (seismograms) produced by such scenario earthquakes is highly beneficial for earthquake engineering considerations. Engineers require site-specific ground-motions for detailed structural response analysis of existing structures and future structures designed around specified performance levels. As a future goal, these scenarios will also provide synthetic time histories of strong ground-motions that include rupture directivity effects.

An example of a ShakeMap scenario earthquake is shown in Figure 2.17 for a hypothetical magnitude 6.9 earthquake on the Newport-Inglewood fault near Los Angeles. Due to the proximity to populated regions of Los Angeles, this scenario represents one the most destructive earthquakes that could impact the region. The U.S. Army Corp of Engineers recently used an event similar to this scenario for evaluating their capacity to respond to such a disaster and to continue to build cooperative relationships with other Federal, State, and local emergency response partners.

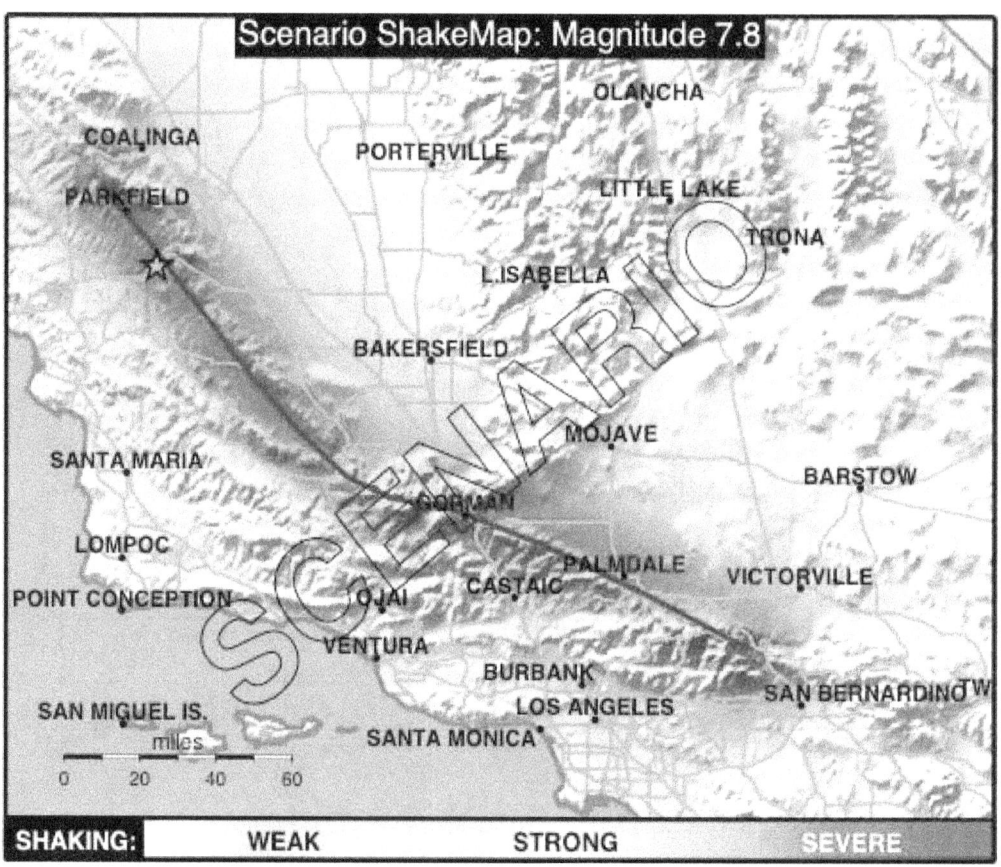

Figure 2.18 Example of a ShakeMap Scenario Earthquake for a hypothetical repeat of the magnitude 7.8 Fort Tejon earthquake on the San Andreas Fault. The format of this map is the TVShakeMap, with larger features suitable for broadcast television resolution.

The next example of a scenario earthquake represents a repeat of the great 1857 Fort Tejon earthquake. The length of the rupture is well established from paleo-seismological studies. This scenario represents a rough estimate of the possible shaking distribution for southern California's "Big One". The scenario, shown in Figure 2.18, is portrayed in the "TV" ShakeMap format, which simplifies the legend for a more general audience as well as accommodates the lower resolution aspects of TV screens compared to computer monitors.

These and other scenarios are available online at the ShakeMap Web pages. They are formatted the same as other ShakeMaps, so they too can be easily used in response planning and loss estimation as well as for educational purposes. They can be found from the *Map Archive* link at the top of all ShakeMap Web pages.

The USGS is planning a concerted effort to promote the use of Scenario earthquake ShakeMaps for all regions of the United States.

2.11 Composite ShakeMaps

Because it's inception as a near-real time data-driven map of shaking distribution, additionally constrained by empirical ground-motion estimates in areas without instrumentation, ShakeMap has been expanded to include other forms of observations and ground-motion predictions. In this section we define our terminology and describe the current range of input constraints and describe examples of the variety of circumstances that warrant specific approaches to combine different post-earthquake data sets. A commonality of all ShakeMaps is the consistent use of gap-filling predictions combined with interpolations corrected for site-specific amplification.

2.11.1 Definitions

ShakeMap.
A near real-time, data driven map with data gaps constrained with empirically-based predictions (attenuation relationships). Once known, fault finiteness is added to the empirical regression to compute distance for the regression more accurately.

Historical ShakeMaps (Major Earthquakes).
Ground-motions constrained with strong-motion observations, typically analog recordings and other with fewer stations than more recent earthquakes. Data gaps are constrained with empirically-based predictions.

Scenario ShakeMaps.
All ground-motions are empirically estimated for a specified fault geometry and a given magnitude. Fault finiteness is included explicitly.

Composite ShakeMaps:
Composite ShakeMaps consist of some combination of observed strong motions and macroseismic intensities, combined with amplitudes estimated from empirical relationships and/or theoretical estimates from forward waveform modeling of finite-fault rupture model. Utilizing macroseismic intensities is accomplished by either using the intensity values as data points on the instrumental intensity maps and by converting an integer Modified Mercalli Intensity (MMI, Wood and Neumann, 1931) or a decimal Community Internet Intensity (CII, Wald and others, 1999c) value into peak ground-motions via inverse of the ground-motion verses intensity relationships of Wald and others (1999b). This is exactly the opposite approach used in

the standard ShakeMap instrumental intensity maps for which ground-motions are related to color-coded intensities via the same relations.

2.11.2   Combining Macroseismic Data with Scenarios

One form of Composite ShakeMap consists of combining macroseismic intensity data with empirical predictions. This is beneficial when historical intensity observation can substantially augment empirical predictions. This is particularly true for very large events for which the empirical relations have few constraining data points, *Example*: 1906 San Francisco, Magnitude 7.9 earthquake (Figure 2.19).

2.11.3   Combining Macroseismic and Instrumental Data

Even for well-instrumented, relatively-populated areas like Silicon Valley of central California, recent earthquake ShakeMaps contain substantial data gaps. However, for the 2002 Gilroy (M4.9), the Community Internet Intensity Maps registered over 17,000 responses, allowing for very detailed and robust intensity observations. These intensity observations can be treated as "stations" and added directly to the instrumental intensity map as observational constraints. Further, by converting these measurements to peak ground-motions amplitudes they provide more detailed images of the contoured ground-motion maps. For areas with few seismic instruments, such observed Macroseismic intensity values can be crucial. *Example*: 2002 Gilroy, M4.9, earthquake (Figure 2.20).

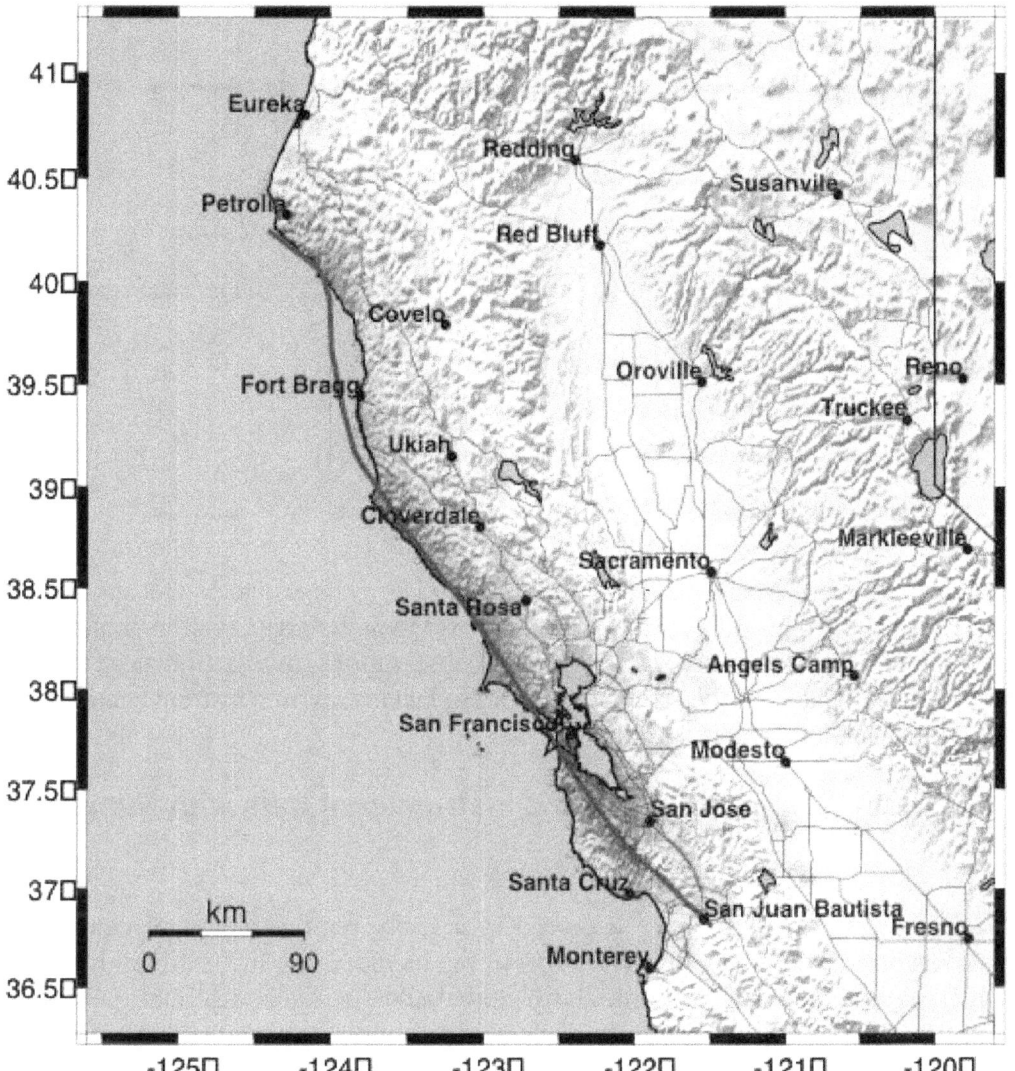

-- Earthquake Planning Scenario --
Rapid Instrumental Intensity Map for San Andreas 1906 Scenario
Scenario Date: Fri Feb 21, 2003 04:00:00 AM PST   M 7.9   N37.77 W122.50   Depth: 0.0km

PLANNING SCENARIO ONLY -- PROCESSED: Mon Apr 14, 2003 01:01:02 PM PDT

PERCEIVED SHAKING	Not felt	Weak	Light	Moderate	Strong	Very strong	Severe	Violent	Extreme
POTENTIAL DAMAGE	none	none	none	Very light	Light	Moderate	Moderate/Heavy	Heavy	Very Heavy
PEAK ACC.(%g)	<.17	.17-1.4	1.4-39	3.9-9.2	9.2-18	18-34	34-65	65-124	>124
PEAK VEL.(cm/s)	<0.1	0.1-1.1	1.1-3.4	3.4-8.1	8.1-16	16-31	31-60	60-116	>116
INSTRUMENTAL INTENSITY	I	II-III	IV	V	VI	VII	VIII	IX	X+

Figure 2.19 Example of a ShakeMap Scenario Earthquake for a hypothetical repeat of the magnitude 7.9 San Francisco earthquake on the San Andreas Fault. Triangles show Modified Mercalli intensity (MMI) observations used as constraints by treating these intensities and associated ground-motions as "data".

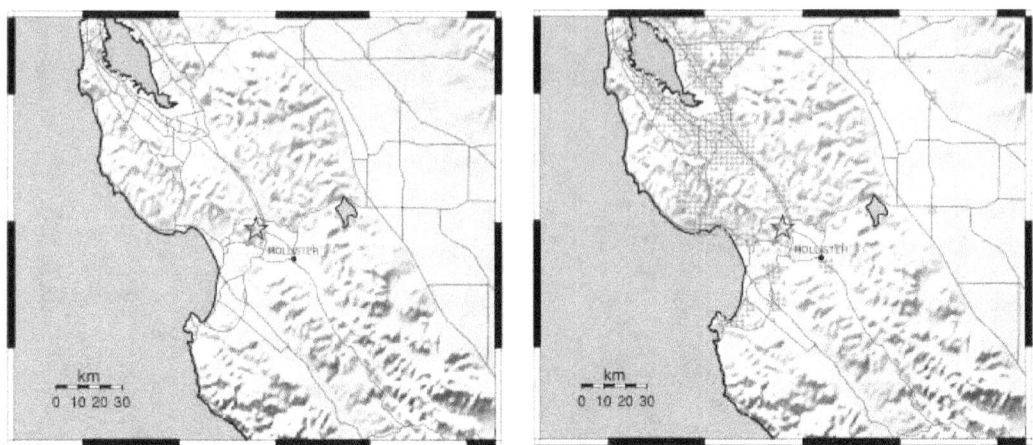

Figure 2.20 Left: ShakeMap for 2002, Magnitude 4.9 Gilroy earthquake, with stations shown as yellow triangles. Right: Combination of strong motion data (yellow triangles) with Community Internet Intensity (CII) intensity observations (orange triangles). The addition of the CII data provides constraints in areas lacking seismic instrumentation; otherwise the maps are similar.

2.11.4   Combining Macroseismic and Instrumental Data with Numerical Predictions

The 2002 Denali (M7.9) earthquake occurred in a fairly remote region of central Alaska. Ground-motion observations were relatively sparse, but included one site nearly right on the fault trace. Other stations were quite distant and included sites in Anchorage and Fairbanks. We augmented these strong motion data with observed intensities at numerous locations both near the fault and throughout the State of Alaska collected with both traditional postal questionnaires and CII values collected online. Finally, we use the finite-fault inversion rupture model of Ji and others (2003) to forward predict peak ground velocities in the near-fault region and combine these predictions with those bias-corrected, empirically-estimated peak motions at greater distances where there are no data (Figure 2.21). The combination of observations and predictions provides a much more complete picture of the distribution of shaking than any of these data sets alone.

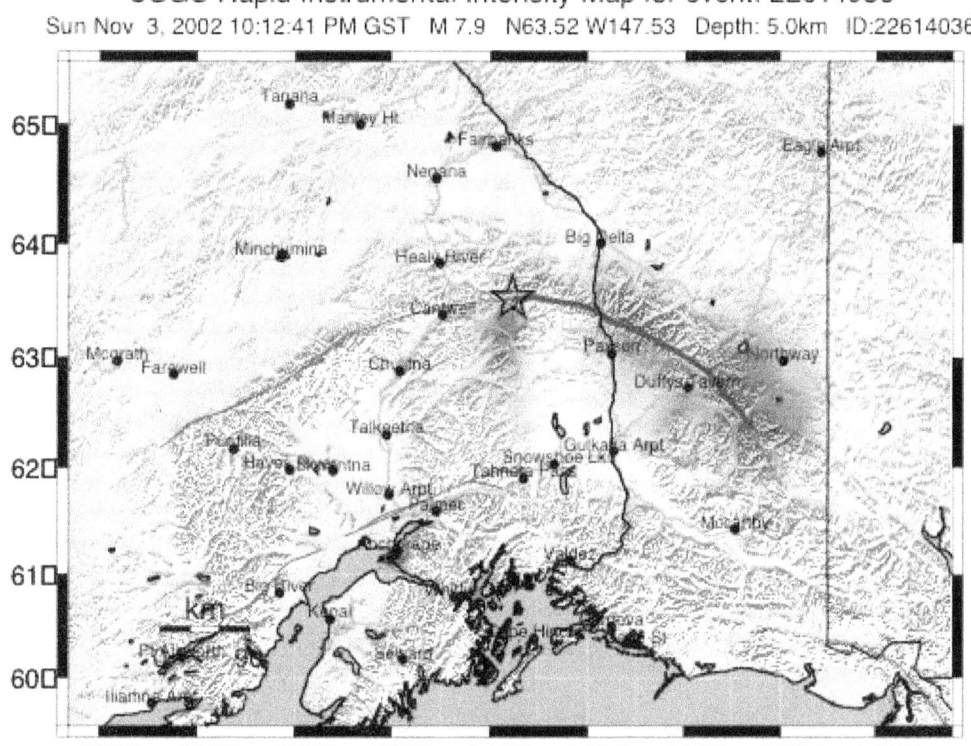

USGS Rapid Instrumental Intensity Map for event: 22614036
Sun Nov 3, 2002 10:12:41 PM GST   M 7.9   N63.52 W147.53   Depth: 5.0km   ID:22614036

PROCESSED: Fri Jan 31. 2003 08:30:04 PM GST.

PERCEIVED SHAKING	Not felt	Weak	Light	Moderate	Strong	Very strong	Severe	Violent	Extreme
POTENTIAL DAMAGE	none	none	none	Very light	Light	Moderate	Moderate/Heavy	Heavy	Very Heavy
PEAK ACC.(%g)	<.17	.17-1.4	1.4-3.9	3.9-9.2	9.2-18	18-34	34-65	65-124	>124
PEAK VEL.(cm/s)	<0.1	0.1-1.1	1.1-3.4	3.4-8.1	8.1-16	16-31	31-60	60-116	>116
INSTRUMENTAL INTENSITY	I	II-III	IV	V	VI	VII	VIII	IX	X+

Figure 2.21 Combination of strong motion data, Community Internet Intensity (CII) intensity observations, and numerical predictions. Most of the near-fault region lacked strong motion recordings so the numerical and CII data are essential. At greater distances the empirical prediction fills in regions without observations.

3 SOFTWARE GUIDE

The following conventions are used throughout this *Guide*:

`Courier Text & prompt (%)`	User Input, commands, and screen displays
< brackets >	User-assigned or environment-specific <Variables>
italics	ShakeMap and non-ShakeMap *programs*
-italics	required or optional program *flag*
'single' or "double quotes"	"file" or "subdirectory" names
http://www.Web.org	Web Page URL

ShakeMap is a collection of programs, largely written in the Perl programming language. These programs are run sequentially to produce ground-motion maps (as PostScript and JPEG images, GIS files, etc.) as well as Web pages and email notifications. In addition to Perl, a number of other software packages are used. In keeping with our development philosophy, all additional software required by ShakeMap is freely available. This chapter explains what is required to install and run ShakeMap.

3.1 System and Software Requirements

Before ShakeMap can be installed and run, a number of other software packages and Perl modules must be installed. This required software is described in the sections that follow.

V3.0: Because ShakeMap V3.0 is substantially different from earlier versions, we have included V3.0-specific notes in set-off paragraphs like this one.

3.1.1 Operating System

V3.0: Support for FreeBSD operating system is new.

ShakeMap was developed and tested on systems running the SPARC version of Solaris V2.6 and V2.7. We have recently completed a port of ShakeMap to the FreeBSD operating system, and this version of ShakeMap (V3.0 and up) will run on FreeBSD. This port allows ShakeMap to be run on inexpensive PC hardware. We do not provide instructions for installing FreeBSD itself, but we have tried to make note of any differences between the Solaris and FreeBSD installations of ShakeMap. We have never tested ShakeMap with the x86 version of Solaris, but we expect that it would work. For any other OS, you will be blazing your own trail. In particular, many of the programs would probably work under another OS, but *transfer* might be problematic. In addition, the makefiles we use are very Unix-like and probably use Solaris-specific extensions (we get around this on FreeBSD by using *gmake*, which supports the extensions we use). Finally, we use SCCS and Teamware for source code control, and it is not at all clear what other platforms are supported. We'll probably switch to CVS at some point, but don't have a timetable for that yet.

3.1.2 Perl

Perl should be installed on any system upon which ShakeMap will run. We are using version 5.005_03, use others at your own risk. (Specifically, we know that Perl 5.8 does not work, so don't try to install ShakeMap with Perl 5.8 unless you want to do the port yourself, which we would appreciate, but couldn't help you with.) Perl may be obtained for free from several sources. Visit www.perl.com to find a download point for your particular OS. You may get the Sun Solaris version on the same FTP site that holds the ShakeMap Source.

We also use several modules that may be obtained from CPAN (see www.cpan.org for CPAN archives). For FreeBSD users, most of these modules are available for automated installation via the ports collection. Modules needed (and recommended order of installation):

V3.0: Modules that should be upgraded from earlier versions are marked with a '+'.

Module Name	Version		
Net::libnet	(1.607)	+	(needs upgrade to 1.16 for ShakeCast)
DBI	(1.13)		
DBD::mysql	(2.1026)	+	
HTML::Template	(2.0)		
XML::Parser	(2.27)		Requires expat be installed*
XML::Writer	(0.3)		
enum	(1.016)		
File::Spec	(0.8)		Built in to later versions of perl (5.6+)
Time-modules	(100.010301)		
Event	(0.78)		
Mail::Sender	(0.7.10)**		
DBD::Oracle	(1.03) ***		
Modules new to ShakeMap 3.0			
Config::General	(2.21)		
MIME::Base64	(2.20)		
URI	(1.24)		
HTML::Tagset	(3.03)		
HTML::Parser	(3.28)		
Digest::MD5	(2.26)		
libwww-perl	(5.69)		
XML::Simple	(2.08)		
No longer needed (for V3.0)			
Text::CSV_XS	(0.20)		
SQL::Statement	(0.1016)		
DBD::CSV	(0.1022)		

*Expat can be downloaded from http://sourceforge.net/projects/expat/. Configuration and installation are explained in the expat README.

**Newer versions of Mail::Sender are available, but they do not work with perl 5.005_03. The later versions use the 'warnings' module (via 'use warnings'), which only comes with newer perl revisions.

***DBD::Oracle is needed to connect to an Oracle database. It is used by programs like db2xml, eq2xml, etc. If you are using a database other than Oracle, you will need to get a different driver (e.g., DBD::Sybase). If you are providing data to ShakeMap through some other mechanism, you won't need this module.

3.1.3 GMT

V3.0: Requires installation or upgrade to GMT 3.4.X. Also, old GMT defaults files should be removed and replaced with ones configured for 3.4.X.

ShakeMap requires GMT, The Generic Mapping Tools developed by Paul Wessel and Walter H.F. Smith. GMT is freely available from http://gmt.soest.hawaii.edu/. We have now upgraded the software to use Version 3.4.X. Use other versions at your own risk as the flags and options are known to change from time to time.

Note: when installing GMT, you will be asked about the type of units used for plotting maps. We use 8 1/2 by 11 (inch) paper, so we have specified all the plot units in inches. You should therefore specify "US" when asked about the type of units. If you end up with very small maps, you probably have specified metric units; change the units to "US" in your GMT defaults file.

If this is not a new install of ShakeMap (i.e., you are upgrading), you will want to remove all of the existing (pre-3.4.X) .gmtdefaults files from the ShakeMap directories, and create new ones.

3.1.4 convert

V3.0: No changes.

Starting with ShakeMap version 2.4 *genex* uses *convert* from ImageMagick to convert PostScript to JPEG. The program can be obtained from www.imagemagick.org. It is free. Ghostscript (see below) is required for *convert* to process PostScript. We are using versions 5.4.2 and 5.4.7 of convert.

3.1.5 PBM/PBMPLUS

PBMPLUS was used in pre-2.4 versions of ShakeMap. It is no longer required.

3.1.6 Ghostscript

V3.0: No changes.

Ghostscript is used by *convert* for conversion of PostScript to JPEG. We use various versions of Aladdin Ghostscript (5.01, 5.50, 6.53). Use whatever version is recommended for your version of convert. The software is free and can be tracked down through the Aladdin Website: www.aladdin.com.

3.1.7 Make

V3.0: GNU *make* is now supported. To make the Solaris and FreeBSD versions work from a common code base, the 'install' program on Solaris has been changed to */usr/ucb/install*. Please double check your 'macros' file in <shake_home>/include after you run *make* in <shake_home>/install to be sure that the correct version of 'install' is selected.

On Solaris, use Sun's make or GNU make (www.gnu.org).

On FreeBSD, you will want to get *gmake*, the GNU make from www.gnu.org. This is easily installed (as are many of the Perl modules) through the ports collection.

3.1.8 SCCS

V3.0: For FreeBSD installations, SCCS may be obtained by installing the cssc (note the clever transposition of characters) package from the ports collection.

SCCS is required for the ShakeMap makefiles to function correctly. SCCS comes with Solaris by default, and may be installed through the cssc package in the ports collection on FreeBSD.

3.1.9 C compiler

V3.0: The CFLAGS macro has been moved to '<shake_home>/include/macros' to allow compilers and compiler flags other than Sun's.

You will need a C compiler. On Solaris, we use Sun's, and on FreeBSD, we use GNU's (again, use the ports collection to install *gcc*). If you will use *gcc* on Solaris, you can get it from (www.gnu.org). In either case, you will set the compiler and compiler flags in '<shake_home>/include/macros.'

3.1.10   MySQL

V3.0: MySQL is new to V3.0.

Please follow the instructions in the section 59951 \h IFigure 2.18}33.1, below, for configuring MySQL, and for converting existing ShakeMap databases to MySQL.

3.1.11   mp (Metadata Parser)

V3.0: Metadata production is new to V3.0.

ShakeMap now produces FGDC-compliant metadata and provides it as text, HTML and XML on the downloads page. Producing the HTML and XML requires the program 'mp' (which should be obtained from http://geology.usgs.gov/tools/metadata/tools/doc/mp.html).   Once you have installed ShakeMap (see installation instructions, below), download, gunzip, and untar the MP software. Cd to the tools/src directory. For Solaris, do the following:

```
% mkdir ../bin
% make -f Makefile.sun all
```

On FreeBSD, copy the file <shake_home>/util/Makefile.bsd to <metadata_home>/tools/src, then do:

```
% make -f Makefile.bsd all
```

In both cases, now cd to <shake_home>/bin and do:

```
% ln -s /path/to/metadata/tools/bin/mp
```

Where "/path/to/metadata" is replaced with the actual path to the directory in which you unpacked the source code, or installed the binaries.

3.1.12   Zip

V3.0: Previously, *zip* was used only to pack the GIS files into archives. With V3.0, zip can also be used to compress the PostScript files and the text grid file to save disk space and reduce transfer times. Zip is still not required, if you do not use these features.

Zip allows the creation of compressed archives. It may be downloaded from www.info-zip.org/pub/infozip (though, again, FreeBSD users can find it in the ports collection). Once you have installed *zip* on your system, there is a configuration parameter *zip* in 'genex.conf' that should be given the full path to the *zip* executable. *Zip* is only required if *genex* is run with either the *–shape* option or the *–zip* option.

3.1.13   Ssh

V3.0: No change.

The secure shell, *ssh*, should be installed if you intend to transfer ShakeMap files via the 'scp' protocol. This is currently required, for example, if you will be transferring your Web pages to the USGS servers. If *ssh* is not available on your system, please see your system administrator – he or she will want to make sure the installation is done correctly and in accordance with your network security policy.

3.2 Installing the Software

3.2.1 Installing and Configuring MySQL

Download MySQL from www.mysql.com. Binary distributions are available for Solaris 8 and 9. If you are using an earlier version of Solaris, you may have to get the source and do a compile or you can get a pre-compiled, though older, version of MySQL from www.sunfreeware.com. If you are using FreeBSD, MySQL is, as usual, found in the ports collection and installation is almost trivial. We are using versions 3.23.53 and 4.0.13, though newer versions will probably work, as well.

We will not describe the MySQL installation process. Extensive documentation is available both online and in the distribution. You will need to get the MySQL server (mysqld) running, and set up an init script to start the server when the machine boots. Be especially careful to follow the instructions for setting a root user password and making sure your MySQL server is secure. You will be asked to do something like:

```
% cd /usr/local/mysql
% ./bin/mysqladmin -u root password 'your_root_password'
```

or:

```
% ./bin/mysql -p
Password:
(give an empty password)
...
mysql> set password for
    -> root@your_machine=PASSWORD('your_root_password');
```

(The following instructions assume that your MySQL server is running on the same machine that you run ShakeMap. This configuration is not required; you may run MySQL on another machine, but you will have to modify some of the commands given below to include a host name. See the MySQL documentation for more information. Also, keep in mind that your ShakeMap system will only be as reliable as the combined reliability of these two machines (i.e., consider providing backup power for both machines, their router, etc.).)

The first step is to create a database and a user. Connect to the MySQL server as root. To connect and be prompted for a password:

```
% mysql -u root -p
Password:
(type your password and hit 'return')
...
```

```
mysql>
```

Now establish the shake database (we call it 'shakemap,' but you can call it anything you want as long as that is the name you use throughout the installation and configuration process):

```
mysql> create database shakemap;
```

Now give the users permission to modify the table. Here we give the user 'shake' (mysql password 'shake_password') the needed permissions:

```
mysql> grant select,insert,update,delete,create,drop,alter
    -> on shakemap.*
    -> to shake@localhost
    -> identified by 'shake_password';
Query OK, 0 rows affected (0.00 sec)
```

Below we have listed the above lines in a format that makes them easy to copy-and-paste into MySQL:

LINES TO CUT-AND-PASTE:
```
grant select,insert,update,delete,create,drop,alter
on shakemap.* to shake@localhost identified by 'shake_password';
```
END LINES TO CUT-AND-PASTE (don't forget to change the password...)

Also create a user 'admin' to do backups:

```
mysql> grant select on shakemap.* to admin@localhost;
Query OK, 0 rows affected (0.00 sec)
```

LINES TO CUT-AND-PASTE:
```
grant select on shakemap.* to admin@localhost;
```
END LINES TO CUT-AND-PASTE

You may wish to create databases for other users, as well. Simply create a separate database for them, and then modify the above command to use the new username and database. For example:

```
mysql> create database jims_database;
...
mysql> grant select,insert,update,delete,create,drop,alter
    -> on jims_database.*
    -> to jim@localhost
    -> identified by 'jims_password';
```

LINES TO CUT-AND-PASTE:
```
grant select,insert,update,delete,create,drop,alter
on   jims_database.*   to   jim@localhost   identified   by
'jims_password';
```

END LINES TO CUT-AND-PASTE (don't forget to change the username and password...)

The other users will have to configure their 'mydb.conf' and 'password' files accordingly, and can then use the included programs to create the tables and convert their old 'shake_flags' and 'earthquake' databases. Note there does not have to be direct correspondence between system usernames and MySQL usernames. Multiple users can share the same MySQL database either through a shared MySQL username, or individual MySQL usernames that all have permission to access the database.

For an explanation of the way ShakeMap uses the database and tables, see the section "**Error! Reference source not found.**" below.

3.2.2 Installation and Upgrade

V3.0: Because this is a major upgrade, we recommend doing a clean install of the software. Existing mapping and data files (e.g., geology, topography, roads) may be copied to the new version without change. Some configuration files have changed substantially, but existing '.conf' files can still be used as guides. You may wish to copy them to the new '<shake_home>/config' directory before executing the final 'make all' command.

To begin, install the software packages and modules described in the section "**Error! Reference source not found.**" above. Stick with the recommended versions, even if they are older and harder to find. If you are upgrading, there are some new modules, and some of the existing modules will need to be updated.

For the installation of ShakeMap you will be making two directory trees: one for the source, <shake_src>, (where you can do development) and another for the online program, <shake_home>, (which you will customize to your environment). Once the directories are created get the ShakeMap source code from ftp.gps.caltech.edu in the directory /pub/shake/src (login as 'anonymous,' or 'ftp'). The file will be named 'shakemap_<major rev>_<minor rev>.tar.gz.' Untar the code in <shake_src>:

```
% cd <shake_src>
% gunzip —c shakemap_3_0.tar.gz | tar xvf —
...
```

Table 3.2A provides a description of each of the top-level directories and Table 3.2B lists some of the more important subdirectories.

Now you will create a version of ShakeMap that is customized for your computing system. To do this (on Solaris):

```
% cd <shake_src>/install
% make
```

On FreeBSD, do:

```
% gmake INSTALL=/usr/bin/install
```

(In the instructions that follow we will use *make*, for which the FreeBSD users should substitute *gmake* unless their GNU make is installed or aliased to 'make.')

Edit the file '<shake_src>/include/macros.' This file sets the paths to some of the required software packages as well as flags for some programs. Next, issue the following commands:

```
% cd <shake_src>
% make dist
```

When this is done, you should have a file '<shake_src>/shake.tar.' Create a directory for the online version of ShakeMap (e.g., /opt/ShakeMap on Solaris or /usr/local/ShakeMap on FreeBSD), which we'll call <shake_home>. Then:

```
% cd <shake_home>
% tar xf <shake_src>/shake.tar
% cd <shake_home>/install
% make
```

Yes, you're doing this last step in two places, but it is needed to make the makefiles work. You will also need to edit <shake_home>/include/macros again. Alternatively, you can copy the file '<shake_src>/include/macros' to '<shake_home>/include/macros.' Then:

```
% cd <shake_home>
% make all
```

Make outputs to the screen any errors and any configuration files that must be edited. Table 3.2C describes some additional top-level directories that will exist following this last step.

The next step in installing ShakeMap is to customize for your specific geographic region. To do this you will need to install a number of data files, and modify the configuration files in the directory '<shake_home>/config.' More information about the customization process can be found in section **Error! Reference source not found.**, **Error! Reference source not found.**; complete the customization process described there before proceeding with this section. (Don't forget to comment out the line "program : scfeed" in 'shake.conf.')

V3.0-specific block:
If this is a new install or upgrade to V3.0, it will be necessary to create tables in the MySQL database. This is easily accomplished:

```
% cd <shake_home>/bin
% ./mktables
```

This process will not destroy the tables if they already exist; to do that, connect to MySQL and issue the proper "drop table" commands. Errors in this program are not usually fatal: if one or more tables already exist, the program will complain, but will continue and make any tables that do not yet exist.

If this is an upgrade to V3.0, you will want to convert the existing earthquake and shake_flags databases to MySQL. Programs exist for this purpose as well. These programs assume that the files '<shake_home>/database/earthquake' and '<shake_home>/database/shake_flags' exist. If they do not (possibly because you are actually following instructions and have installed this version of ShakeMap in a new directory), simply copy them from their old location into the new <shake_home>/database . Do the following:

```
% cd <shake_home>/bin
% ./eq2mysql
% ./shake2mysql
```

These programs will complain if the data they are inserting already exists, so if you need to correct errors, first drop (and recreate (with *mktable*)) or truncate the tables before running the programs again. Once you are satisfied with the results (as determined by running an event and looking at the home and archive pages on your web site), you will never use these programs again. It is unlikely that this will all work perfectly the first time. Feel free to run the programs, edit your 'earthquake' and 'shake_flags' files, drop and recreate the tables, and run the programs until it all works. Nothing will break. A simple way to check your work is to connect to MySQL and have a look at the table:

```
mysql> use shakemap;
...
mysql> select * from earthquake order by tabsol;
```

(You will want a nice, wide window to view this information.) This will display all of your archived events in chronological order (or use 'evid' instead of 'tabsol,' above, to see events ordered by event id).
End of V3.0-specific block.

Once the config files have been edited, the final step for installation is to create the web products and put them on the web server. To do this:

```
% cd <shake_home>/lib
% make web
% cd <shake_home>/bin
% ./transfer -permweb
```

Check that the transfer was successful. You will probably need to run and transfer an event before the web pages will work properly.

V3.0: Because V3.0 introduces compression of web products and a dramatically more efficient directory structure (both within the local 'data' directory and on the web sites), you may wish to rerun many (or all) of your existing events to save space. You will also want to delete all of the events from your web site(s). If you wish to do this but minimize the down time of the site, you can make a dummy web site on a local machine and modify 'transfer.conf' to transfer only there. Then rerun all of your events. Finally delete the existing web site(s) and copy the dummy site to the web server(s) (and don't forget to change 'transfer.conf' back to its original configuration). You could accomplish the same thing by omitting *transfer* from the processing of each event, then deleting the events from the web site, then running *transfer* for all the events in sequence. Our web sites ended up being about 40% of their original size when we performed this task.

Note that within the ShakeMap <shake_home> directory the subdirectory 'data' will contain all the event data and intermediate files as well as the final products to be transferred. Depending on the number of events, and the resolution of your grid and topography files, this directory can grow to be quite large. If disk space is limited on the install partition, the 'data' directory should be placed on a larger partition and a link to it (called 'data') should be made from the install directory. E.g.:

```
% cd $SM_HOME
% rmdir data
% ln -s /bigdisk/shake_data data
```

3.3 Customizing ShakeMap

3.3.1 Region-Specific Files

There are a number of region-specific files that you will need to create (see Table 3.2A and Table 3.2B). You should give these files names different from those in the distribution or they will be overwritten when you upgrade. Most of these files are part of the configuration defined in 'mapping.conf' and 'grind.conf.' See the configuration files themselves for more documentation.

3.3.2 Configuration Files

In the directory <shake_home>/config you will find a number of configuration files. It is important to read the documentation within these files as they provide most of the information necessary to customize ShakeMap to your particular environment. Table 3.2C lists the ShakeMap programs and the configuration files upon which they depend. All of the programs also depend on 'mydb.conf' to access the MySQL database. More discussion of shake.conf and mysql.conf can be found in the section "Running ShakeMap."

When editing configuration files, please note that the default values (as described in the documentation for some parameters) may not be the same as the value assigned to the parameter by default within the configuration file itself. The assigned value is the recommended value, the documented default is only used if no assignment is made, and may no longer be the recommended value (but may have been retained for reasons of backward compatibility).

Important Note: When editing shake.conf, please comment out the line:

```
program : scfeed
```

The program 'scfeed' will not function until a ShakeCast server is generally available and your system is configured to connect to it.

(When upgrading please note: From time to time we make changes to programs that require changes to config files. These changes must be merged with the config files that the user may have modified in customizing his/her version of ShakeMap. This is a non-trivial problem, and our solution is a bit simplistic. The merging consists of inserting the user's potentially changed config statements as comments into the new config file. The user may then go through the file and select which config statements are appropriate. This process takes a few minutes, but is fairly easy. Except in the case of 'transfer.conf,' which turns into a mess when it is changed. In this case it is often easier to clean out the destinations and file lists in the new config, then go to the backup file 'transfer.conf.BAK' (always made to keep a safe copy of the user-modified config files around) and just cut and paste your old destinations and file lists back into the new config file.)

3.3.3 Passwords

You will need passwords to access a database through db.conf or mydb.conf (or for *transfer* using ssh or ftp). To set up a password file:

```
% cd <shake_home>
% mkdir pw
% chmod og-rx pw
% cd pw
```

Create or copy your passwords file to 'passwords.' For an explanation of the format of this file, see '<shake_home>/src/lib/Password.pm.' Also see the section "Running ShakeMap," below for more on 'mydb.conf.' In general, the format for *ssh* and *FTP* passwords is:

```
<machine> <username> <password>
```

And for database access the format is:

```
<dbname> <username> <password>
```

where the substitutions for "dbname" and "username" above should exactly match the strings in the database configuration file.

3.3.4 Web Pages

You may also wish to make changes to the Web pages. We have tried to include much of the region-specific data in the Web.conf file, but there may be additional customizations needed. Please keep track of your changes and let us know so that we can add common items to the configuration file. The Web pages and templates can be found in <shake_home>/lib/genex/Web/.

3.3.5 Automation

Because each regional network is different, automation is left to you. Currently code exists to automate generating ShakeMaps from two types of systems: 1) a database running the NCEDC/SCEDC schema (as in southern California and Berkeley), and 2) *earthworm* running with the Oracle database. If you are using either of these systems you will be able to adapt current code.

If you do not use one of the above data acquisition systems, you will need to first generate code that will process data in near-real-time. The output of this processing should include peak horizontal acceleration, peak horizontal velocity, and 5 percent-damped peak horizontal acceleration (0.3, 1.0 and 3.0 second periods) for all horizontal component data. This information along with station information must be written into ShakeMap compatible XML files with filenames that end in "_dat.xml." The event information – latitude, longitude, depth, and magnitude – should be written to a second ShakeMap compatible XML file – "event.xml". See the section on "ShakeMap XML Input," below, for a discussion of these file formats. Examples of data and event XML files can be found in the distribution in the directory <shake_home>/data/9583161/input.

Next, you need a program to watch when these files are made, then copy them to the ShakeMap input directory and start ShakeMap. This could, of course, be the same program that creates the files.

The distribution includes a program called 'queue' and its associated configuration file 'queue.conf' that may be of interest. *queue* waits for an alarm announcing an event or cancellation (see the programs 'shake_alarm' and 'shake_cancel') and then takes appropriate action depending on its configuration (i.e., given a location and magnitude it will either kick off a run of ShakeMap or ignore the event). It can prioritize and queue multiple events, and schedule events for automatic reprocessing at user-defined intervals. The program accesses a database to retrieve information on the earthquake, but should be fairly easy to adapt to other systems.

If you develop a program (or modify *queue*) that you think might be of interest to other ShakeMap installations, please let us know and we will include it in a future release.

3.3.6 Attenuation Relations

V3.0: The calling convention for maximum() and random() has changed. Please be sure to update your custom modules to reflect this change. See the example modules (e.g., <shake_home>/src/lib/Regression/Small.pm) for examples of the new calling convention.

Custom attenuation relations may be needed for some regions. If you are going to develop a module, the interface must be modeled after the ones found in <shake_src>/src/lib/Regression (e.g., Small.pm).   The module should also be added to the file "<shake_src>/src/lib/Regressions.pm."

3.4 Running ShakeMap

ShakeMap consists of a series of programs (refer to list Table 3.2) that when run sequentially, produce the desired output and transfer it to its destination. All of the programs will print documentation when run with the '*-help*' flag, and most of them have an associated configuration file (found in the "config" directory and named "<program>.conf") that controls the behavior of the program.

3.4.1 Data Directory Structure

Before running ShakeMap you must collect some data. This data is stored in the data directory, and as mentioned elsewhere, it can become quite large. Put it somewhere with lots of space and link to it from your distribution directory. Each event is stored in its own sub-directory named for the event, whether this be a number or a text string. This event name must be the same as in the file containing the event information – "event.xml". Within each event directory a number of subdirectories are created (Table 3.4). ShakeMap will create all of these directories except "raw" and "input".

3.4.2 Creating the Maps

Once the ShakeMap software is installed and configured, creating a ShakeMap is simple. First, cd to <shake_home>/bin (e.g. /opt/ShakeMap/bin), then execute 'shake':

```
% ./shake -event <event_id>
```

This will run the pre-configured set of programs as specified in "shake.conf". If you would like a little more information about the progress of the run, use the *-verbose* flag to 'shake'.

It is not always appropriate or necessary to run all of the programs. For instance, when running a historic event, or an event not otherwise in the database, the 'retrieve' program will probably fail, causing 'shake' to abort. One possibility is to reconfigure "shake.conf" to skip the unnecessary program(s). Another option is to use the *-dryrun* flag:

```
% ./shake -event <event_id> -dryrun
```

Which will produce output showing the programs that shake would run (and their options) without actually running them:

```
/opt/ShakeMap/bin/retrieve -event 9108645
/opt/ShakeMap/bin/pending -event 9108645
/opt/ShakeMap/bin/grind -event 9108645 -qtm -boundcheck
      -lonspan 4.5 -psa
/opt/ShakeMap/bin/mapping -event 9108645 -timestamp -ascii
/opt/ShakeMap/bin/shakemail -event 9108645
```

```
/opt/ShakeMap/bin/tag -event 9108645 -mainshock
/opt/ShakeMap/bin/genex -event 9108645
/opt/ShakeMap/bin/print -event 9108645
/opt/ShakeMap/bin/transfer -event 9108645 -www -ftp
```

You may then run the programs you choose and ignore the others. For instance, if you were to make a change to the "estimates.xml" file, you might just run 'grind' and 'mapping' and then look at the plots as PostScript (the .ps files in the "<shake_home>/data/<event_id>/mapping" directory). You could then run 'genex' and look at the JPEGs. Or also run 'transfer' and look at the images on your Web site.

3.4.3 The Gory Details

Of course, it is never that simple. And even if it were, there are reasons for having a better understanding of the system. Here, then, is more detailed information on configuring 'shake' and on the way the versioning system works.

3.4.3.1 shake.conf

The program 'shake' is the main ShakeMap program. Its job is to run a series of other programs in a specified order, possibly calling the programs with invocation flags that vary with magnitude. The program can also be told to call certain programs only the first time a given event is processed. Run *shake -help* to see other options.

At this point, it is recommended that you read 'shake.conf' (in '<shake_home>/config') to get a basic idea of what is available. The default configuration is probably about right for most installations (except for the "program : scfeed" line, which you will want to comment out until you are configured to communicate with a ShakeCast server). Some of the parameters ('once_only,' 'no_dep,' 'cancel,' and 'scenario_skip') probably won't need to be changed unless you add a new program to the processing sequence with the 'program' parameter (and maybe not even then).

'shake.conf' is also the configuration file for the program 'cancel,' which effectively undoes the effects of *shake*, removing the event from the system, sending cancellation notices, and rebuilding the web pages to reflect the absence of the cancelled event.

3.4.3.2 The Processing Sequence and shake.conf

ShakeMaps are not always automatically generated. Frequently, manual intervention is necessary or desirable, and we often run one or more of the programs repeatedly until we are satisfied with the results. For example, the automatic processing sequence might go something like this:

> *retrieve* → *pending* → *grind* → *tag* → *mapping* → *genex* → *shakemail* → *transfer* → *setversion* → *scfeed*

But after the automatic run, we might wish to change the map dimensions or centering by changing the options to *grind*. Our manual sequence might look like this:

grind → *mapping* → *genex* → *transfer* → *scfeed*

We might run the *grind* → *mapping* pair several times in succession until we are satisfied with the results. Satisfied, we then run *transfer* to update the web pages with our new maps. Previous versions of ShakeMap would happily do this, despite the fact that we forgot to run *genex* and, as a result, some of our products (e.g., the PostScript maps) do not agree with others (e.g., the JPEG maps and shapefiles).

Starting with ShakeMap V3.0 we have introduced the idea of program dependency. Simply put, a program is considered to be dependent on the programs that precede it in the processing sequence, and it will not run unless the sequence is run in the proper order. For instance, in the above example, *transfer* would recognize that *mapping* had run more recently than *genex* and would abort with an error message explaining the problem.

Things to be aware of:
1) The processing sequence is defined by the order of 'program' lines in 'shake.conf.'
2) A program that does not affect the performance of programs later in the sequence (i.e., later programs do not depend on its output) can be identified with a 'no_dep' line in 'shake.conf.' For instance, *shakemail* sends email to interested parties, but does not generate data that any program later in the processing sequence depends upon. Thus, *shakemail* is declared 'no_dep.' When a later program (e.g., *transfer*) runs, it will not include *shakemail* in its investigation of the processing sequence. But (this is important!) *shakemail* itself will still require the programs that precede it to be run in sequence. Thus, if *shakemail* is run immediately after *mapping*, it will complain that *genex* has not been run.
3) You do not have to always start at the beginning of the sequence. Once an event has been run once, you can start anywhere in the sequence. You can jump in and re-run *mapping*. You can run it a bunch of times in a row. Then you can run *genex*. Then you can run *mapping* again. Then you can run *grind*. What you can't do is use out of date output.
4) Yes, it seems complicated. But it is actually simple. Assume the function T() returns the time a program, P, was most recently run. Assume that 'P_n' is the nth non-no_dep program in the processing sequence. The software enforces the relation:
$$T(P_1) < T(P_2) < \ldots < T(P_{n-1})$$
with the provision that each of the n-1 earlier programs has run at least once.
5) You can always force a program to run with the *-forcerun* flag.

So how does the system keep track of all this? By using the 'shake_runs' database table described in the next section.

3.4.3.3 Flags, Versions, and the MySQL Database

During the ShakeMap installation process you created a number of tables in your MySQL database. These tables replace the old 'earthquake' and 'shake_flags' Text::CSV tables in pre-

V3.0 ShakeMap, and provide functionality to support versions and the processing sequence integrity system described above.

The database tables in the shakemap database can be listed with *mysql*:

```
mysql> use shakemap;
Database changed
mysql> show tables;
+--------------------+
| Tables_in_shakemap |
+--------------------+
| earthquake         |
| server             |
| shake_lock         |
| shake_runs         |
| shake_version      |
+--------------------+
5 rows in set (0.00 sec)
```

The 'server' table contains information the ShakeCast system needs to connect to a server. This information will be provided to individual regions when the ShakeCast system is fully available (early 2004 is the target date).

The 'earthquake' table is very similar to the earlier CSV table of the same name:

```
mysql> describe earthquake;
+-----------+-----------+------+-----+---------+-------+
| Field     | Type      | Null | Key | Default | Extra |
+-----------+-----------+------+-----+---------+-------+
| evid      | char(80)  |      | PRI |         |       |
| name      | char(255) | YES  |     | NULL    |       |
| locstring | char(255) | YES  |     | NULL    |       |
| tabsol    | datetime  | YES  |     | NULL    |       |
| tzone     | char(8)   | YES  |     | NULL    |       |
| mag       | double    | YES  |     | NULL    |       |
| lat       | double    | YES  |     | NULL    |       |
| lon       | double    | YES  |     | NULL    |       |
| mainshock | char(20)  | YES  |     | NULL    |       |
| cluster   | char(80)  | YES  |     | NULL    |       |
+-----------+-----------+------+-----+---------+-------+
10 rows in set (0.00 sec)
```

This table is accessed and modified by a number of programs (*tag*, *genex*, *cancel*, etc.). Its primary purpose is to maintain a complete inventory of the events for which ShakeMaps have been made. Under rare circumstances you may have to edit this table (using SQL commands), so the following table describes the columns.

Name	Description	Valid values
evid	The event identifier.	Any text string that forms a valid Unix filename, up to 80 characters.
name	The long, possibly descriptive name of the event; will be printed at the top of the maps.	Any text string up to 255 characters.
locstring	The location of the earthquake. If the name field is not specified (through the program 'tag'), this text will be used as the event name on the maps.	Any text string up to 255 characters.
tabsol	The date and time of the event in the format: yyyy-mm-dd hh:mm:ss	From 1000-01-01 12:00:00 AM to 9999-12-31 11:59:59 PM
tzone	The timezone of 'tabsol,' above.	Usually 'GMT,' but could be 'PST,' 'MDT,' etc.
mag	The earthquake magnitude.	Any valid magnitude.
lat	The latitude of the earthquake epicenter.	North is positive, south is negative.
lon	The longitude of the earthquake epicenter	West is negative.
mainshock	Value set by the program 'tag' to categorize the earthquake	Valid values include '', 'current,' 'historic,' 'scenario,' and 'invisible.'
cluster	If this event is part of a larger sequence, this field specifies the evid of the mainshock in the sequence. This may be useful for creating a special archive page for a particular sequence.	Any valid evid.

The table 'shake_lock' table is used to prevent multiple ShakeMap processes from operating on an event at the same time. Each ShakeMap program will acquire the lock before it begins processing, and will release the lock when it quits (or is killed).

```
mysql> describe shake_lock;
+---------+----------+------+-----+---------+-------+
| Field   | Type     | Null | Key | Default | Extra |
+---------+----------+------+-----+---------+-------+
| evid    | char(80) |      | PRI |         |       |
| program | char(80) |      |     |         |       |
| pid     | int(11)  |      |     | 0       |       |
| tepoch  | int(11)  |      |     | 0       |       |
+---------+----------+------+-----+---------+-------+
4 rows in set (0.00 sec)
```

The columns are: the event id, the name of the program, the process id of the locking process, and the Unix epoch time that the lock was acquired. Occasionally, a lock will be held when the locking process is dead or hung. The lock can be broken by 1) using the '-forcerun' flag to the next program, or 2) calling the program 'unlock' with the event id of the locked event (this program will also optionally try to kill the locking process), or 3) if a lock is stale (more than fifteen minutes old), ShakeMap programs will automatically unlock the event and continue processing after issuing a warning message.

The 'shake_runs' table keeps track of the last run of each program for each version of an event. But first:

A Digression on Versioning

After a great deal of discussion and consideration, we decided that the most useful demarcation of a 'version' of a ShakeMap (which is really a collection of products) is the point at which the products are distributed to external destinations. In other words, we create a new version every time we run *transfer*, whether or not that version differs in any significant way from the previous version. (Models that assigned version numbers to each product based on its difference from the previous version of that product, while sexy, were ultimately found to be too complicated, unreliable, and unworkable. Consider, for example, a JPEG map that varies in no way from another map, except that the embedded processing date is different. Is that a different version? Some say "yes," some say "no." Plus, no one could come up with a compelling reason for defining versions this way. But our digression digresses…)

So how does this versioning system work? Let us assume that *transfer* has just run on an event and created version 'N' (if *transfer* has never run for this event, 'N' would be zero). We then run one of the other programs in the processing sequence. For instance, we run *grind* to change the "lonspan." The program will inspect the 'shake_version' table and determine that the most recent version of the event is version 'N.' *grind* will then declare itself to be working on version 'N+1.' It will check that the processing sequence is being honored, do its processing job, then insert some information about itself (its name, the current time and date, the version, and the flags with which it was invoked) in the 'shake_runs' table before exiting. If we were to run this program again, it would go through the same process, but when it found that a row already existed in the shake_flags table for that event/program/version combination, it would simply update the date/time and invocation flags. It would still be version N+1. We could run it twenty times and it would still be version N+1. We could then run *mapping* (version N+1) and *genex* (version N+1). We could go back and run *grind* some more (still version N+1). Finally, when we run *transfer*, the new version is declared complete, a new row is inserted in 'shake_version' for version N+1, and the products are transferred to the world. The next time a program in the sequence is run, it begins version N+2. And so on.

(In the situation where some programs were not run, the missing programs are inserted into the 'shake_runs' table with the new version number, but the date/time and flags of the previous version. For example, we could run *mapping*, *genex*, and *transfer*, without ever re-running *grind* (which is a valid thing to do – see the section on the Processing Sequence, above). When the new version was set, the system would copy the flags and time/date of the previous run of *grind*, but give it the new version number.)

By using this system, we have a complete record of the programs and their invocation flags for each version of the event that we transferred to the world. In conjunction with the judicious use of the program 'setversion' (which will save a copy of the input data and the configuration files for an event in a version-specific directory) we can recreate any version of an event. Here is the a listing of a southern California event:

```
mysql> select program,flags from shake_runs where
evid='14007388' and version=4 order by lastrun;
+----------+---------------------------------------------------+
| program  | flags                                             |
+----------+---------------------------------------------------+
| retrieve |                                                   |
| grind    | -qtm -boundcheck                                  |
| mapping  | -timestamp -notchecked -plotests -tvmap -itopo    |
| genex    | -zip -metadata -shape shape                       |
| transfer | -www -ftp -push                                   |
| scfeed   |                                                   |
+----------+---------------------------------------------------+
6 rows in set (0.01 sec)
```

By running these programs, with these flags, on the preserved input data and the preserved configuration files, we could re-create version 4 of this event.

Keep in mind:
1) *transfer* sets a new version unless you tell it not to with *-noversion*.
2) Versions can be created by *setversion*. *setversion* will also delete, modify, or query the version information for an event.
3) The default invocation of *setversion* (i.e., "*setversion –event* <event_id>") does nothing. Use the magnitude-dependent flags in 'shake.conf' to configure *setversion* to save the data for significant events without filling your disks up with data from a lot of magnitude 3.5 earthquakes.
4) *transfer* has a *-forget* flag that will prevent its flags from being saved in the database. This is useful for *cancel*, and *pending*, or if you are doing something unorthodox. *grind* also has a *-forget* flag. All of the programs probably should.

End of Digression

The 'shake_flags' table has the following structure:

```
mysql> describe shake_runs;
+----------+------------+------+-----+---------+-------+
| Field    | Type       | Null | Key | Default | Extra |
+----------+------------+------+-----+---------+-------+
| evid     | char(80)   |      | PRI |         |       |
| program  | char(80)   |      | PRI |         |       |
| lastrun  | datetime   | YES  |     | NULL    |       |
| version  | int(11)    |      | PRI | 0       |       |
| flags    | char(255)  |      |     |         |       |
+----------+------------+------+-----+---------+-------+
5 rows in set (0.00 sec)
```

Most of the columns are self-explanatory: the event id, the program name, the date/time of the last run, the version, and the invoking flags (sans the '-*event* <event_id>' and '-*verbose*' flags). Note that the primary key consists of (evid, program, version).

Version information is stored in the 'shake_version' table:

```
mysql> describe shake_version;
+----------+-----------+------+-----+---------+----------------+
| Field    | Type      | Null | Key | Default | Extra          |
+----------+-----------+------+-----+---------+----------------+
| evid     | char(80)  |      | PRI |         |                |
| version  | int(11)   |      | PRI | NULL    | auto_increment |
| lddate   | datetime  | YES  |     | NULL    |                |
| comment  | char(255) | YES  |     | NULL    |                |
+----------+-----------+------+-----+---------+----------------+
4 rows in set (0.00 sec)
```

The columns are obvious except for 'comment.' If the version was created by *transfer*, the comment will be "Automatic call from within transfer." If you use *setversion* to make the version, you can give a comment on the command line.

3.4.3.4 Passwords and mydb.conf

The configuration line for MySQL access in mydb.conf will look something like this:

```
database : mysql shakemap shake password
```

where you would substitute your database name for 'shakemap' and the username of the user running ShakeMap for 'shake.' E.g., 'jims_database' and 'jim' if user jim is experimenting with his own version of ShakeMap. See the section "Installing and Configuring MySQL" for instructions on giving jim his own database. If you are running MySQL on a remote machine, your config line will look something like this:

```
database : mysql database=shakemap;host=machine.domain.org
shake password
```

In the password file ('<shake_home>/pw/passwords,' by default), you will need a line:

```
shakemap shake <mysql_password_for_user_shake>
```

or, if you are using a remote database server:

```
database=shakemap;host=machine.domain.org shake
<mysql_password_for_user_shake>
```

with the obvious substitutions to make it work in your environment (or jim's). Yes, the "database=shakemap...)" bit looks wrong, but the password module is comparing strings with what is found in "mydb.conf" and this is what is required to make it work.

3.4.3.5 Backing up the MySQL database

Because we are maintaining a database, and because what we keep in our database is important, it is probably a good idea to do database backups on a regular basis. There are a number of ways to do this with MySQL, including logging every transaction in a way that lets you recreate the database after any failure. See the MySQL documentation for details if you would like to implement a more robust backup system than is described here.

The *mysqldump* program allows one to dump one's tables to a file as a set of SQL statements that can then be used to recreate the tables. For example:

```
% mysqldump --add-drop-table -u admin shakemap >
shakemap.sql
```

The file so created can then be used to restore the database (or to transfer the data to another system):

```
% mysql -u shake -p shakemap < shakemap.sql
Password:
```

Note that the user names and database name may need to be changed on your system. Also note that for *mysqldump* we use the 'admin' user that we created in the section "Installing and Configuring MySQL". This user does not need a password because its only SQL permission is SELECT.

We have included a program 'mysqlbu' in the directory <shake_home>/util. This program performs the database dump, compresses the output and, optionally, copies the output to another machine for safekeeping. (The program contains hard-wired path and machine names, though, so you will have to ~~hack~~ modify it for your system.) 'mysqlbu' can be run daily – it will create a different file for each weekday. The program also prints an error summary that can be piped to a mail program. We run it with a crontab entry that looks like this:

```
0 2 * * * /home/shake/bin/mysqlbu | mail —t shake_admin
```

Which runs *mysqlbu* at 2:00 AM every day, and mails the status report to the user 'shake_admin.'

3.4.4 A Note about Shake Flags

Because ShakeMaps are often generated (or regenerated) automatically, there needs to be some way to preserve manual modifications. For instance, a certain event is run by the queue, and then the operators decide that the scale should be larger, so they run the event manually, using the -latspan flag to grind. If this information were not preserved, any subsequent automatic run of that event would revert to the original settings. Thus, we created the "shake_flags" database, which keeps track of the parameters with which each program was last run. The program 'shake' and ONLY the program 'shake' (this is important) reads that database and uses the flags found there when running each of the subprograms.

This can result in confusing behavior. For instance, if you were to make some changes to the Web pages for a particular event, and then run transfer with only the *-www* flag (because only Web changes were made), the next run of shake on that event would run transfer with only the -www flag, which would not update the ftp site, which might lead to confusion.
Because transfer is often used this way, it has the -forget flag, which effectively prevents it from updating the shake_flags database for that run. 'shake' has the *-default_fl* flag which causes shake to ignore the "shake_flags" database and use the default flags for each sub-program as specified in the config file.

Keep this in mind when you are manually running events. You have been warned.

3.4.5 A Note about CSV Databases

The "shake_flags" and "earthquake" databases are currently implemented as CSV (comma-separated value) databases through the DBD::CSV *PERL* module. This implementation has the advantage of being simple and fast and the files can be manually edited (if you're very careful). It has the huge disadvantage of being totally at the mercy of program and system errors. Killing a program with Control-C can screw up your entire database. We will probably replace this system with a big heavyweight database like *MySQL* or *Postgres*, which is total overkill, but which provide some degree of transaction safety and data integrity. In the meantime, you should back up the earthquake and "shake_flags" databases (found in the "database" directory) periodically.

3.4.6 A Note about Estimates and Flagged Stations

'grind', unless directed otherwise, will attempt to make estimates of ground-motion (based on an attenuation relation of your choosing) and will flag (i.e. cause not to be included in the maps) stations that appear to be outliers. It will put these estimates and flagged stations into files in the "ShakeMap/data/<event_id>/richter" directory. If a file called "estimates.xml" exists in the "ShakeMap/data/<event_id>/input" directory, these estimates will be used instead of those produced automatically by 'grind' (but 'grind' will still compute the estimates for the purpose of flagging outliers). If a file "flagged_stations.txt" is in the "input" directory, it will be used in preference to the one computed by 'grind'. Thus, if "estimates.xml" and "flagged_stations.txt" are in the "input" directory, ;grind' will use them, and not compute its own.

So, if you compute estimates via some external program and place them in the input directory, grind will use them, but will flag outliers based on its own model. If you are using a sophisticated slip distribution model, you probably want to compute your own outliers and put them in a file "flagged_stations.txt" in the "input" directory, too.

Finally, unless "estimates.xml" and "flagged_stations.txt" are in the input directory, grind will always recompute the estimates and outliers. The files in "richter" are regenerated with each run. We do this because the input data could change (e.g. additional data arrives or the event magnitude is revised), and the estimates should reflect this fact.

3.4.7 A Note about Finite Faults

Events now accept an optional finite fault file that will be used in generating estimates (for real events or scenarios), and can be plotted on the map using the xyaddon feature in "mapping.conf". The filename must end in "_fault.txt" and should be placed in the event's input subdirectory.

The finite fault file is composed of a set of (latitude, longitude) points defining the surface trace of a fault. For example, two points can define a simple strike slip fault. A closed polygon (first and last points identical) can represent a dipping fault. **NOTE**: The reverse order of the points ((latitude, longitude) or (y,x) rather than (x,y) is an unfortunately legacy format that would be difficult to correct given the number of ShakeMap scenarios already in existence.

ShakeMap computes distance-to-fault to each line segment in the fault and uses the closest distance. A point inside a closed polygon is considered to be at zero distance. Note that the default ShakeMap regression computes Joyner-Boore distance (to the surface projection of the fault), so fault depth is ignored.

The file should be formatted as the input of the GMT 'psxyz command' (a '>' header, followed by space-delimited lon-lat pairs.)

3.4.8 Sending Email

There are two options for sending email. One uses the program 'shakemail' to send a text message notifying the user group that a ShakeMap has been made, details about the source, and a link to the Webpage. Two uses the program 'shakemail_attach'. This program sends the above text message, but it also attaches a JPEG version of the intensity map. 'shakemail_attach' must be run after 'genex'.

3.4.9 Scenarios

ShakeMap now supports the generation of earthquake scenarios. The user need only create the appropriate *_dat.xml, event.xml, and (optionally) "estimates.xml: and finite fault files (see item 1, above) in an input directory. The scenarios are distinguished from real earthquakes in one of two ways: A) through the conscientious use of the -scenario flag in the many programs (not recommended, or B) by ending the event id with "_se" (e.g. <SHAKE_HOME>/data/ myscenario_se/input) (highly recommended).

Scenario earthquakes are distinguished from real ones by a truly stunning number of appearances of the word "Scenario" on the maps and Web pages, including a big one emblazoned across the face of the maps themselves. We do this to prevent the misunderstandings in the press and public that would surely occur if we were any less zealous. Trust us. Scenarios have their own place on the archive page, distinct from the real earthquakes, and they will not appear in the real event lists or on the homepage.

Most of the programs are now scenario-savvy. 'Shakemail', for instance will not email scenarios unless you force it to. 'Transfer' will transfer to Web sites (*-www*) and ftp sites (*-ftp*) but will not push (-push) unless you force it to. Run the various programs with *-help* to see the new scenario-related options and behavior.

To create a new scenario, the most straightforward way is:
1) Create a new event subdirectory (say, "data/1857_se") and a new "input/" directory under that ('data/1857_se/input').
2) Copy the "event.xml" file from an existing event over to the new input directory, and modify the parameters. (Don't forget to change the 'id' field.)
3) Add a finite fault file, if desired (see Finite Faults, above.)
4) In the file "database/shake_flags", add a line describing your new scenario. Most of the flags for scenarios are optional, except for the '-scenario [scenario-description]' in the 'tag' field.
5) Run 'shake *-event* <1857_se> *-dryrun*' just to make sure all the flags are correct. Then run it without '*-dryrun*'.

Note: Because the estimate grid for a scenario is much finer than the usual (non-scenario) grid and requires lots of computation, ShakeMap will compute the grid once and store it for future use. Use the 'grind' *-forcests* flag to recompute the estimate grid (when changing a regression parameter, for example.)

3.5 Common Problems

We welcome contributions to this section. Please let us know about problems you have had with ShakeMap, and your workarounds (if any).

3.5.1 Shake flags database causes confusion

See "A Note about Shake Flags," above.

3.5.2 Files in incorrect format

When configuring region-specific files, make sure to create files following the formats in the example (i.e., southern California) files. If the code is written to read a space-delimited file, commas will cause problems and vice versa. For the *GMT* files make sure you have the latitude and longitude in the correct columns.

3.6 XML Formats in ShakeMap

3.6.1 About XML

XML is a system for tagging text to indicate the structure of information in the text. XML started as a generalization of HTML (or a simplification of SGML, depending on your perspective), and XML markup is similar in appearance to HTML tags. However, in XML the tags are defined on a per-application basis. With this flexibility, XML can be used as a means of structuring data in a cross-platform, human-readable form, in addition to its use handling textual documents.

A complete specification of XML is available at http://www.w3.org/TR/REC-xml (http://www.w3.org/TR has a number of interesting documents) and an annotated version is at http://www.xml.com/axml/axml.html.

However, preparing XML files for ShakeMap does not require knowing the specification. For working with ShakeMap, it will probably be enough to get a short summary, in particular contrasting XML with the more familiar HTML.

An XML file starts with a declaration line:

```
<?xml version="1.0" encoding="US-ASCII" standalone="yes"?>
```

Version refers to the XML standard to which the file is written. Currently, "1.0" is the only option. Encoding refers to the character set in which the file is written. Standalone indicates whether the XML file is free of references to outside definitions in other XML files.

Following the declaration is an optional Document Type Definition (DTD) block, which may refer to a definition in another file:

```
<!DOCTYPE earthquake SYSTEM "earthquake.dtd">
```

or present the definition in place:

```
<!DOCTYPE earthquake [
  ... DTD description ...
]>
```

Then the XML itself starts. XML tags look a lot like HTML tags with a tag label and possibly attributes:

```
<tag att1="val1" att2="val2">
```

In contrast to HTML, XML tags and attributes are case sensitive, so <station> and <STATION> are different. Also, attribute values must always be wrapped in quotes, so <station code="PAS"> rather than <station code=PAS>.

In HTML, some tags are simple tags that don't contain other tags or blocks of text. For example:

```
<img src="..." border="0">
```

The equivalent in XML is called an empty tag, and only differs from HTML by closing with /> rather than >:

```
<pga value="0.25"/>
```

Non-empty tags contain blocks of other tags and/or character data, such as:

```
<station code="PAS">
 <comp name="HLN">
  <acc value="0.25"/>
 </comp>
</station>
```

Example codes that demonstrate writing XML are available in the ShakeMap distribution package (in <shake_home>/src/xml), and because XML files are text files this consists mainly of simple printing of formatted output. For input, XML parsers are freely downloadable for the Perl, C and Java programming languages. ShakeMap is predominantly written in Perl, so we use a well-regarded parser library in that language. As with XML output, example codes in the ShakeMap distribution show how input parsing is handled. A list of XML parser libraries in various programming languages is available at http://www.w3.org/XML/#software.

Every XML file has a set of tags used in a pattern particular to that type of file. This pattern is set by the developer and can be indicated in a Document Type Definition (DTD). The DTD defines the tags that it expects, the order it expects them in, and how tags can nest within one another. It also indicates what tags are optional, what tags can appear multiple times in succession, what attributes are associated with each tag, and (optionally) a range of values accepted for an attribute. There is also a concept of an XML schema, but we will not go into that here.

Some parsers have an option to 'validate' an XML file according to its DTD, but the parser used by ShakeMap does not yet do so. However, we have found it useful to define DTD's for the various XML file types that ShakeMap works with, if only for documentation purposes during development. These ShakeMap DTD's will be discussed below for each file type.

3.6.2 ShakeMap XML Files

Before ShakeMap is run for a particular event (identified by an event id), the following set up is needed:
- o a directory in <shake_home>/data/<event_id>/input
- o an 'event.xml' file in this directory
- o one or more files with filenames ending in '_dat.xml' in this directory

The contents of the 'event.xml' file are earthquake parameters in the 'earthquake.dtd' format. This format is a single, empty tag with a number of attributes of the earthquake. The attributes are given in the following table.

Event information	
id	the event id
created	file creation time (Unix epoch -- seconds because Jan 1, 1970)
Hypocenter information	
lat	latitude (in decimal degrees, negative in southern hemisphere)
lon	longitude (in decimal degrees, negative in western hemisphere)
depth	in km, positive down
locstring	a free-form descriptive string of location relative to landmarks
mag	magnitude
Origin time parameters	
year	4 digit format
month	1-12
day	1-31
hour	0-23
minute	0-59
second	0-59
timezone	abbreviation (i.e., GMT, PST, PDT)
Amplitudes at the epicenter	
pga	peak acceleration (units of %g)
pgv	peak velocity (units of cm/s)
sp03	Spectral acceleration at 0.3 sec period (units of %g)
sp10	Spectral acceleration at 1.0 sec period (units of %g)
sp30	Spectral acceleration at 3.0 sec period (units of %g)

As mentioned, the amplitude attributes in 'earthquake.dtd' are estimates produced by ShakeMap during processing. These attributes should be left out of the 'event.xml' file input to ShakeMap, and will be ignored if present.

An example 'event.xml' file look like:

```
<?xml version="1.0" encoding="US-ASCII" standalone="yes"?>
<!DOCTYPE earthquake [
  ... DTD description ...
]>
<earthquake id="14000376" lat="34.2722" lon="-118.7530"
mag="3.6" year="2003" month="10" day="29" hour="23" minute="44"
second="48" timezone="GMT" depth="13.81" locstring="2.6 mi W of
Simi Valley, CA" created="1069292035" />
```

Files in the input directory named like '*_dat.xml' are station parameters in the 'stationlist.dtd' format. This format has a root 'stationlist' element containing one or more 'station' elements. The 'stationlist' can have a 'created' attribute with the file creation date in Unix epoch time (seconds because Jan 1, 1970). Each station has a set of attributes:

code	the station code
name	station name and/or description
insttype	description of instrument type
lat	station latitude (in decimal degrees)
lon	station longitude (with negative sign in western hemisphere)
source	agency that maintains the station (i.e., SCSN, CDMG, NSMP,...)
commtype	digital or analog communications (DIG or ANA)
loc	free form text describing the location of the station (optional)

Each station element contains one or more 'comp' elements. Comp elements have the following attributes:

name	a channel name/code in SEED convention
originalname	the original channel name if it was not SEED (optional)

The name attribute must be a SEED-convention name. If the name is not known, for example if the source of amplitudes only gives a single summary value for the station, then use the most generic code for a horizontal component, HL1. Use a horizontal code rather than HLZ because ShakeMap uses only horizontal components in processing.

If the amplitude is from an agency that does not use SEED component codes, you will have to map their codes to a comparable SEED code for the name attribute. If you would like the original code carried through the processing and used in the HTML, XML and text stationlists, then put the original code in the originalname attribute.

Each 'comp' element must contain one 'acc' element, and one 'vel' element, and may contain 'psa03,' 'psa10,' and 'psa30' elements (one of each). These refer to peak acceleration, velocity, and pseudo-spectral acceleration (at 0.3, 1.0, and 3.0 sec period) values for the named channel at the named station. The acc, vel, psa03, psa10, and psa30 elements are empty but have the following attributes:

value	the amplitude value
flag	flag indicating problematic data (optional)

The value attributes are expected to have units of:

acc	%g
vel	cm/s
psa	%g

The flag attribute indicates problematic data. Any value other than "0" (zero) or "" will cause ShakeMap to reject the amplitude (and, in fact, all the amplitudes of that type for that station).

ShakeMap also does automatic flagging of outliers (see the program *grind* and the section "Running ShakeMap," above, for more information on automatic flagging). Though any non-zero flag will kill an amplitude, the following flags are currently defined:

T	Automatically flagged by ShakeMap as an outlier
M	Manually flagged (in grind.conf) by the ShakeMap operator
G	Amplitude clipped or below the instrument noise threshold
I	Incomplete (a data gap existed in the time window used to calculate the amplitude)

An example of a '*_dat.xml' file is:

```
<?xml version="1.0" encoding="UTF-8" standalone="yes"?>
<!DOCTYPE stationlist [
  ... DTD description ...
]>
<stationlist created="1070030689">
<station code="ADO" name="Adelanto Receiving Station"
insttype="TriNet" lat="34.55046" lon="-117.43391" source="SCSN
and TriNet" commtype="DIG" loc="Adelanto, on Hwy 395  ">
<comp name="HHE">
<acc value="0.0083" flag="0" />
<vel value="0.0030" flag="0" />
<psa03 value="0.0146" flag="0" />
<psa10 value="0.0049" flag="0" />
<psa30 value="0.0003" flag="0" />
</comp>
<comp name="HHN">
<acc value="0.0088" flag="0" />
<vel value="0.0028" flag="0" />
<psa03 value="0.0111" flag="0" />
<psa10 value="0.0040" flag="0" />
<psa30 value="0.0004" flag="0" />
</comp>
<comp name="HHZ">
<acc value="0.0087" flag="0" />
<vel value="0.0016" flag="0" />
<psa03 value="0.0080" flag="0" />
<psa10 value="0.0013" flag="0" />
<psa30 value="0.0002" flag="0" />
</comp>
</station>
... additional station tags ...
<station code="WSS" name="West Side Station" insttype="TriNet"
lat="34.1717" lon="-118.64971" source="SCSN and TriNet"
commtype="DIG" loc="Hidden Hills, Valley Circle Dr.">
<comp name="HHE">
<acc value="0.0225" flag="0" />
<vel value="0.0031" flag="0" />
```

```
<psa03 value="0.0182" flag="0" />
<psa10 value="0.0016" flag="0" />
<psa30 value="0.0002" flag="0" />
</comp>
<comp name="HHN">
<acc value="0.0209" flag="0" />
<vel value="0.0029" flag="0" />
<psa03 value="0.0234" flag="0" />
<psa10 value="0.0019" flag="0" />
<psa30 value="0.0001" flag="0" />
</comp>
<comp name="HHZ">
<acc value="0.0187" flag="0" />
<vel value="0.0020" flag="0" />
<psa03 value="0.0073" flag="0" />
<psa10 value="0.0005" flag="0" />
<psa30 value="0.0000" flag="0" />
</comp>
</station>
</stationlist>
```

The earthquake and stationlist XML files are combined in the output file provided to the public. This file is made available as XML and is also the basis for a raw, non-XML text stationlist and the HTML Web stationlist linked to the ShakeMap click-maps. Because the output XML file combines the event and station files, it also merges the earthquake and stationlist DTD's into a 'shakemap_data' DTD that is included in the file.

3.6.3 Retrieving Data from a Database

As run by SCSN/TriNet, ShakeMap is triggered by a real-time processing system and accesses a database for event parameters and amplitude values from Caltech/USGS-Pasadena stations. Additional amplitude values are received from CGS and NSMP stations and are incorporated in the processing as they arrive. See the section "External Data XML Files," below.

To access the database, ShakeMap launches *retrieve* which launches any number of specific helper codes (defined in a configuration file) to build the "event.xml" and "*_dat.xml files." These codes can be used as examples of database access to build input files. If your network is running a DBMS with the schemas used by the southern or northern California Earthquake Data Centers, then you may be able to use the ShakeMap codes directly. If you are using a DBMS with a different schema, it will be necessary to modify at least the SQL calls embedded within the example programs and possibly the logic of the programs themselves if the schema differences are large.

3.6.4 External Data XML Files

External (i.e., not directly from database) amplitudes can be included in ShakeMap once they are associated with an earthquake. Just make a stationlist.dtd-format XML file with a unique name ending in _dat.xml and drop it in the correct <event id>/input directory.

In order to associate amps, data need to be received in a structured way. One possibility is defining an XML format. We have taken this approach with CGS (was CDMG) and NSMP data, and the XML format is described here as an example.

CGS (and NSMP) data is sent to ShakeMap in the unassociated data XML format. The main difference between the stationlist XML files fed directly to ShakeMap and the CGS amplitude XML files is the addition of timing information (the basis for the association). The root element of a CGS amplitudes file is an 'amplitudes' element. 'amplitudes' has an 'agency' attribute so we can know who the amplitude report is from. The amplitudes element contains one or more 'record' elements. The record element can have an agency-defined 'id' attribute assigned to it.

The record element contains 'timing' and 'station' elements. The timing element has no attributes but contains 'reference' and 'trigger' elements. The reference element has two attributes, 'zone' for a time zone code (i.e., GMT, PST, or PDT) and 'quality' for an agency-defined indicator of the timing quality. 'reference' contains a set of elements:

year	4-digit year
month	1-12
day	1-31
hour	0-23
minute	0-59
second	0-59 (60 for leap second)
msec	0-999

each of which has an integer 'value' attribute as defined above. 'trigger' is an empty tag with a 'value' attribute assigned the time in seconds of the amplitude trigger relative to the reference time. CGS has a common trigger time for all components in a record, so the trigger tag is not stored at the component level.

The 'station' element has four attributes:

code	station code
name	station name or description
lat	station latitude (in decimal degrees, negative in the southern hemisphere)
lon	station longitude (in decimal degrees, negative in the western hemisphere)

and contains one or more 'component' elements. Each component has a 'name' attribute that defines the component (in an agency-defined way), and contains 'acc', 'vel', and 'sa' elements. Each of these elements has 'value' and 'units' attributes, where value is the amplitude value itself, and units is a string expressing the units (i.e., g, or %g, or cm/s/s). 'sa' has an additional attribute, 'period', that defines the period, in seconds, of the spectral value. For each component, there is one acc, one, vel, and zero or more sa elements.

An example of a CGS amplitude XML file is:

```
<?xml version="1.0" encoding="US-ASCII" standalone="yes"?>
<amplitudes agency="CDMG">
 <record>
  <timing>
   <reference zone="GMT" quality="0.5">
    <year value="2000"/>
    <month value="02"/>
    <day value="21"/>
    <hour value=" 13"/>
    <minute value="49"/>
    <second value="0"/>
    <msec value="0"/>
   </reference>
   <trigger value="0"/>
  </timing>
  <station code="23920" lat="34.004" lon="-117.058"
name="Yucaipa Valley">
   <component name="Up">
    <acc value=" .013" units="g"/>
    <vel value="   .32" units="cm/s"/>
    <sa period="0.3" value="0.01160" units="g"/>
    <sa period="1.0" value="0.00204" units="g"/>
    <sa period="3.0" value="0.00070" units="g"/>
   </component>
   <component name="90">
    <acc value=" .026" units="g"/>
    <vel value="   .63" units="cm/s"/>
    <sa period="0.3" value="0.02261" units="g"/>
    <sa period="1.0" value="0.00418" units="g"/>
    <sa period="3.0" value="0.00135" units="g"/>
   </component>
   <component name="360">
    <acc value=" .028" units="g"/>
    <vel value="   .58" units="cm/s"/>
    <sa period="0.3" value="0.02152" units="g"/>
    <sa period="1.0" value="0.00375" units="g"/>
    <sa period="3.0" value="0.00205" units="g"/>
   </component>
  </station>
 </record>
</amplitudes>
```

Example codes that parse this XML format and convert it to the ShakeMap input format are part of the 'dirwatch' modules found in <shake_home>/src/watcherlib and <shake_home>/src/cdmglib. In particular, see the module watcherlib/AssocAmp.pm.

3.7 Development Model

We are going to try to handle ShakeMap development as an open-source project. This means that various developers will contribute to the project the code that they feel improves the overall product. This also means that those contributions must not be site-specific unless they are easily bypassed by other users (through configuration options, for example). Changes, improvements, additions, etc. will be sent back to Bruce Worden, to be included in the distribution product (or to be sent back to the source for revision). If all goes smoothly, your site may make extensive changes to the core product, send them back to the distribution source, have them integrated into the code base, and then receive them back with the next release of the source. This should lead to (relatively) painless upgrades, not to mention a better product for everyone.

None of this prevents a site from taking the code and running totally wild with it. It simply means that their work will not be included in future releases and upgrades to the core ShakeMap product.

We have elected to use Sun's *TeamWare* as our development environment. In a nutshell this product allows multiple developers to work within their own independent workspace, and to merge their work into a higher-level workspace. This is handled through a parent-child workspace environment:

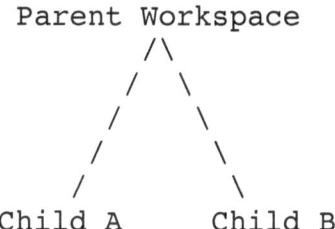

```
          Parent Workspace
                /\
              /    \
            /        \
          /            \
        /                \
      Child A        Child B
```

Developer #1 works in the workspace "Child A" and Developer #2 works within "Child B". (Note: although they may work independently of one another, it is best that they communicate so that they do not work at cross purposes, or even modify the same files too extensively, because this requires a "merging" step that is facilitated by *TeamWare*, but which can be complicated.) When Developer #1 is finished with some development, he does a "putback" to the parent workspace. When Developer #2 then tries to do a putback, he will find that he must first do a "bringover" of the modified parent to his child. As part of this bringover he must reconcile any differences that exist between his work and that of Developer #1. Once he has done this and tested the program, he may complete his putback to the parent.

We strongly suggest you follow this model, even if you only have one developer. The reason is that it will facilitate your returning code to us, and us sending updates to you. Imagine you are working in Child Workspace A, and we send you an upgrade. You can set up this code as Child B and do a putback to the parent (which may require a bringover, as discussed above, if you have previously put back changes to the parent). Once you have done this step, you can putback your latest changes to the parent (which will definitely require a bringover because we know the parent has changed). Similarly, when you have completed development that you believe should be included in the distribution, you can send us the parent directory, and we can merge it into our code in the same way.

All of this depends on you having *TeamWare*. *TeamWare* usually comes with Sun's *WorkShop* product, which you probably bought if you have any of the compilers and debuggers. Older versions are not Y2K savvy, so if you get a bunch of SCCS errors you need to upgrade.

Note that within his own workspace each developer will be working with SCCS commands to check out, modify, and check in individual files. We strongly recommend sticking to this SCCS regimen even if you don't have *TeamWare* because, again, it will facilitate our incorporation of your code into our code base.

3.8 Tables

Table 3.1A. Files and directories in the top-level of ShakeMap

Makefile	The highest-level makefile in the distribution.
config	Initially contains only a README file explaining how the configuration files are formatted; once a 'make' is done, the directory will be populated with various config files for ShakeMap; these files will be edited by the user to conform with the site requirements.
doc	Most of the ShakeMap documentation.
install	The first stop when doing an install of a ShakeMap distribution; see "Installing the Software" above.
lib	Contains Perl modules, mapping and data files, site correction data, Web pages, Web page templates, and supporting graphics; see "Configuring ShakeMap," above, and sections below for more information.
sc	The directory holding the ShakeCast software.
src	The directory where the ShakeMap source code lives.
util	Directory containing a couple of handy programs.
Codemgr_wsdata	This directory contains information used by the "TeamWare" code development tool. See "Development Model" for more about TeamWare.
SCCS	Directory containing data for the SCCS Source Code Control System. Again, see "Development Model" for an explanation of SCCS and how it relates to TeamWare.
deleted_files	Used by TeamWare to store files that have been removed from the distribution; you can safely ignore this directory.

Table 3.1B. Subdirectories of Interest

src/cdmglib	Contains perl modules that are used by dirwatch, the directory watching program; these modules are used in the conversion of CGS XML or CGS two-line parametric files into ShakeMap XML.

src/cfgsrc	The source for the default configuration files; the installation process copies these into <SHAKE_HOME>/config, then merges them with any existing config files. The user then customizes them for a specific environment.
src/config	Contains the modules ShakeConfig.pm and WatcherConfig.pm which hold global variables used by most of the ShakeMap programs; these modules have site-specific customizations made to them and are installed in <SHAKE_HOME>/lib by the program 'config' (also found in this directory). No user intervention is required.
src/contour	Contains the source to the 'contour' program. 'contour' converts GMT .grd files (in the #1 (binary) format) into GIS shapefiles (polygons of "constant" parametric value).
src/genexlib	Directory with modules specific to the program *genex*.
src/lib	Directory containing modules used by several of the ShakeMap programs; most of these modules have (non-POD) documentation within them.
src/misc	Contains a couple of helpful programs: a perl version of 'echo' and the infamous configconfig, the new programs required by the MySQL conversion (mktables, eq2mysql, and shake2mysql), and some other ad hoc programs.
src/queue	Contains the event queueing and automatic ShakeMap initiating program used by the southern California network; individual sites will probably want some custom variation of this program; see src/cfgsrc/queue.conf for customization options; directory also contains the alarming and cancellation scripts.
src/shake	Contains the core of the ShakeMap software; most of these programs have a configuration file (in src/cfgsrc) that explains how each may be customized; see "Shake Programs" below for a discussion of the individual programs.
src/util	This directory holds programs to convert the ascii lat-lon-velocity file to binary and back to ascii; see the section on configuring ShakeMap for more information. Also in this directory are programs to create the instrumental intensity scales for the II map and the TV map.
src/watcher	Contains the dirwatch program; the dirwatch program and its associated modules provide the service of watching a directory for the arrival of a file, and then dispatching that file to its proper destination; see the description of the modules in src/watcherlib, below; see the README in src/watcher for a discussion of the program's capabilities.
src/watchercfg	Contains configuration files for the watcher modules.
src/watcherlib	Currently contains two modules (three, actually, but Base.pm is general purpose): AmpDir.pm: Takes the 2-line CDMG text parameter files as input, converts them to unassociated XML, and deposits the new file in a user-specified directory. AssocAmp.pm: Takes the unassociated XML file, tries to associate it with a TriNet event, converts the XML to ShakeMap XML, deposits this file in the input directory for ShakeMap, and, after waiting a user-specified time, alarms the

	queue that the event has been updated.
src/xml	Contains various programs for converting data files and database results into ShakeMap XML files: eq2xml Probes the TriNet database for information specific to a numbered event then writes an XML file in the event input directory describing the event. db2xml Queries the TriNet database for event-specific amplitudes then writes the appropriate XML. <various> The other programs read various text file formats and generate XML following the stationlist.dtd. This directory also contains the DTD files describing the earthquake and stationlist XML formats.
lib/genex	A collection of HTML and templates that, through the magic of the *genex* program, become the Web site.
lib/mapping	Contains data files used by the *mapping* program: highways, faults, cities, topography, colormap, etc. Much of the customization of ShakeMap happens in this directory. See config/mapping.conf for more details.
lib/ps	Contains the PostScript of the Instrumental Intensity scales for the intensity map and the TV map.
lib/sitecorr	Contains the station velocity file, the site amplification table, and the text and binary versions of the geology file; review these files and create versions specific to your region.
lib/transfer	Contains dummy files used by *transfer* when pushing data files to remote sites.
lib/xml	Holds the DTD's for the ShakeMap XML; the DTD's are prepended to the earthquake and stationlist data files.

Table 3.1C Directories Created After Installation

database	Holds the 'earthquake' and 'shake_flags' databases; discussed below. Now obsolete.
bin	All of the executable programs will end up here after a 'make all.'
data	Repository of all event data and processed files. Discussed below ("Data Directory Structure") and elsewhere.
pw	(Actually, the name and location of this directory is user-defined); this is where database passwords are kept; should be read protected for security; see the db.conf configuration file and the Password.pm module (in src/lib) for examples of use.
perl	Directory where the various perl modules end up after a 'make all'; it is also permissible to install other perl modules used by ShakeMap (e.g. DBI) here.
include	Holds the macros used by makefiles and the config program.

logs	Directory in which the queue puts its logging and error files.
watcher	Host directory where the various directory watcher modules (dirwatch program) look for config files and dump bits of information. May also hold the logs. This directory can be ignored if you do not use the dirwatch program.

Table 3.2A. Region-Specific Files in 'grind.conf'

Parameter: none **File:** lib/sitecorr/ [region]_vsgrid.txt	Geology file. dx by dy (where dx=dy) rectangular grid of the Vs30 values for the ShakeMap region. This file must be comma delimited: lon, lat, Vs30 (where west longitude is negative)
Parameter: qtm_file **File:** lib/sitecorr/ [region]_vsgrid.bin	Binary form of the above file. To generate, run qtmlatlon2bin with above file as input. This must be done on a machine with the same byte order as the ShakeMap machine.
Parameter: ampfactor_file **File:** lib/sitecorr/ site_corr_[region].dat	File containing site amplification factors as a function of Vs30 and frequency of input ground-motion. See the southern California file site_corr_cdmg.dat for documentation.
Parameter: stavel_file **File:** lib/sitecorr/ dig_[region].txt	File containing station information: lat, lon, sta name, Vs30; stations not found in this file will be assigned the Vs30 of the nearest grid point from the geology file, above. This may be the same file that is given as fwstatlist, below.
Parameter: fwstatlist **File:** lib/grind/ [region]statlist.txt	List of stations used by the -*scenario* option (to *grind*) to create dig_dat.xml

Table 3.2B. Region-Specific Files in 'mapping.conf'

Parameter: topo_cmap **File:** lib/mapping/ [region]_elev.cpt	GMT colormap file for plotting regional topography; the default file 'tan.cpt' may work for many regions.
Parameter: map_roads **File:** lib/mapping/ [region]_roads.xy	GMT file containing coordinates of road segments: lon, lat pairs grouped by segment , segments separated by a '>'.
Parameter: map_faults **File:** lib/mapping/ [region]_faults.xy	GMT file containing coordinates of fault segments: lon, lat pairs grouped by segment, segments separated by a '>'.
Parameter: map_topo and map_topo_hires **File:** lib/mapping/ [region]_topo.grd	GMT grid file for the regional topography. Optionally, you can have both high and low resolution forms.
Parameter: topo_intensity and topo_intensity_hires **File:** lib/mapping/ [region]_topo_intens.grd	GMT grid file of intensity for the regional topography grid given above. If this file (or the high resolution version) does not exist, the *mapping* program will generate it.
Parameter: map_cities,	Files containing city names and locations. These files are now

map_bigcities, and map_verybigcities **File:** lib/mapping/ [region]_cities.txt, [region]_bigcities.txt , and [region]_verybigcities.txt	deprecated; use the '_label' versions instead. See 'mapping.conf' for more details. A program 'fix_cities' is provided to convert old city files to new ones; read the program source for documentation.
Parameter: none **File:** lib/mapping/ tvguide.txt	Optional, edit this file to reflect local contact information.

Table 3.2C. ShakeMap Programs

shake	Config: **shake.conf** The main program; actually a wrapper program that calls the other programs. The configuration file controls what programs shake calls and how they are called. After shake calls the first program in the list (usually *retrieve*, see below), it expects a file, "event.xml," in the event's input directory.
retrieve	Config: **retrieve.conf** Usually the first program called by *shake*; retrieve is itself a wrapper code that calls other programs that are meant to retrieve data and put it in the event's input directory; the configuration file explains the customization options.
pending	Sends a new home page to the Web site to indicate that an event is being processed; pending calls *genex* with the *-pending* flag, and *transfer*.
grind	Config: **grind.conf** *grind* reads the data files it finds in the event's input directory and generates grid files with interpolated ground-motions, as well as the text parameter file, and the station and estimate files. *grind* puts its output in a directory called '<shake_home>/data/<event_id>/output.'
tag	ShakeMap keeps an earthquake database that it uses to generate the home page and the archive pages; *tag* specifies to the database that an event is a) ordinary, b) a mainshock, c) an historic, named, event, d) invisible, or e) part of an aftershock cluster associated with a mainshock.
mapping	Config: **mapping.conf, colors.conf** Reads the grids generated by *grind* and makes PostScript maps of ground-motion and shaking intensity, contour files, and generates information needed to make image maps; all of this output is placed in the event's 'mapping' directory.
asciimap	Called by *mapping* (if invoked with the *-ascii* flag); generates the ASCII version of the intensity map; this program is currently southern California specific; it will probably disappear from the next release.
genex	Config: **genex.conf, Web.conf** Uses the output of *grind* and *mapping* to create JPEGs, build Web pages, and generate GIS and other files for export via the Web or FTP.
shakemail	Config: **shakemail.conf** Generates a number of different email notifications of ShakeMap availability (long format, short format, attached JPEG, and list of flagged stations). See

	shakemail.conf for details.
addon	Config: **addon.conf** Creates and copies a QDDS-formatted file to a local QDDS directory; QDDS should then add a link to the just-created ShakeMap from the Simpson maps. Will also send a delete message for cancelled events.
print	Config: **print.conf** Sends plots to printers.
transfer	Config: **transfer.conf** Transfers the output created by *genex* to the Web and ftp sites, also 'pushes' ShakeMap data to remote sites via FTP. *transfer* has been pirated for other uses as well: it is used to transfer the permanent parts of the Web pages to the Web site(s), and it transfers a temporary 'pending' page to the Web while an event is being processed.
setversion	Manipulates the version information for an event and preserves versions as requested. Run *setversion –help* for more information. Also, see the section on version control in this manual.
scfeed	Config: **addon.conf** (to obtain source network code) Creates XML files for an event and its associated ShakeMap products, and calls ShakeCast programs to insert the files as messages into the ShakeCast system. The ShakeCast config file is found in '<shake_home>/sc/conf/sc.conf.'
cancel	Config: **shake.conf** *cancel* undoes the effect of *shake*: it removes the event (except what is found in the input directory) from the data directory and removes the event from the earthquake database; it removes the Web pages for the event and updates the home and archive pages to reflect the removal of the event; it deletes all associated data from the ftp site(s) and it pushes a file, '<event_id>.cancel,' to push clients
unlock	If an event is locked, preventing the execution of ShakeMap programs, this program will break the lock.

Table 3.3 Subdirectories Found Within an Event Data Directory

input	Directory in which the input XML is placed. The operator may also manually transfer estimates and flagged station files into this directory.
output	Directory in which *grind* places its output.
richter	Another directory that contains output from *grind*. The estimate grid and flagged stations files are written here if *grind* is called upon to generate them.
mapping	This directory will contain PostScript files generated by *mapping*, and JPEG files converted from the PostScript by *genex*; also contains contour files, the ASCII map, and other miscellaneous products.
genex	This directory contains products ready for transfer to the Web and ftp sites. It contains two sub-directories 'Web' and 'ftp.' Each of these contains files set up in a directory structure that lends itself to being copied wholesale to its destination.
Raw	This directory is not created by the ShakeMap software, but may be created by

	the user; it is a holding area for input data that is not in the proper XML format. Some programs (*dig2xml*, *ana2xml*, *hist2xml*, etc.) look in this directory for event-specific input which they convert to XML and place in the 'input' directory

REFERENCES

Abrahamson, N.A. (2000). Effects of rupture directivity on probabilistic seismic hazard analysis, *Proc. of 6th Int. Conf. on Seismic Zonation*, Palm Springs, Earthquake Engineering Research Institute.

Abrahamson, N. A., and K. M. Shedlock (1997). Overview, *Seismological Research Letters*, **68**, 9-23.

Ashland, F.X. (2001). Site-response characterization for implementing ShakeMap in northern Utah, Utah Geological Survey Report of Investigation – 248, 10 pp.

Atkinson, G.M. and D.M. Boore (2003). Empirical ground-motion relations for subduction regions and their application to Cascadia and other regions. Bull. Seism. Soc. Am, 93, 1703-1729.

Atkinson, G. M., and D. M. Boore (1997). Some comparisons between Recent ground-motion relations, *Seismological Research Letters*, **68**, 24-40.

Atkinson, G. M., and D. M. Boore (1995). Ground motion relations for eastern North America, *Bulletin of the Seismological Society of America*, **85**, 17-30.

Atkinson, G. (1993). Source spectra for earthquakes in eastern North America. Bull. Seism. Soc. Am., **83**, 1778-1798.

Applied Technology Council (2002). ATC-54: Guidelines for using strong-motion data and ShakeMaps in Post-Earthquake Response.

Applied Technology Council (1985). *Earthquake Damage Evaluation Data for California*, ATC-13 Report, Applied Technology Council, Redwood City, California, 492 pages.

Applied Technology Council (1989). *Procedures for Postearthquake Safety Evaluation of Buildings*, ATC-20 Report, Applied Technology Council, Redwood City California.

Applied Technology Council (1991). *Seismic Vulnerability and Impact of Disruption of Lifelines in the Coterminous United States*, ATC-25 Report, Applied Technology Council, Redwood City, California, 440 pages.

Bazzurro, P. and Cornell, C.A. (2002). Vector-Valued Probabilistic Seismic Hazard Analysis (VPSHA), *Proceedings 7th U.S. National Conference on Earthquake Engineering*, Boston, MA, July, 2002.

Bauer, R.A., J. Kiefer, and N. Hester (2001). Soil amplification maps for estimating earthquake ground motions in the Central US, *Engineering Geology*, **62**, 7-17.

Bauer, R.A. Compilation of databases and map preparation for regional and local seismic zonation studies in the CUSEC region: Collaborative research - Organization of CUSEC State Geologist with assistance from USGS and administrative support from CUSEC. CD ROM.

Bazzurro, P. and Cornell, C.A. (2002). Vector-Valued Probabilistic Seismic Hazard Analysis (VPSHA), *Proceedings 7th U.S. National Conference on Earthquake Engineering*, Boston, MA, July, 2002.

Beresnev, I. A., and K.-L. Wen (1996). Nonlinear soil response - a reality? (A review), *Bull. Seism. Soc. Am.,* 86, 1964-1978.

Boatwright, J., H. Bundock, J. Luetgert, L. Seekins, L. Gee and P. Lombard (2003). The dependence of PGA and PGV on distance and magnitude inferred from Northern California ShakeMap data. *Bull. Seism. Soc. Am.,* 93, no. 5, 2043-2055.

Boatwright, J., K. Thywissen, and L. Seekins (2001). Correlation of ground-motion and intensity for the January 17, 1994, Northridge, California earthquake, *Bull. Seism. Soc. Am.* 91, 739-752.

Boore, D. M., W. B. Joyner, and T.E. Fumal (1997). Equations for Estimating Horizontal Response Spectra and Peak Accelerations from Western North American Earthquakes: A Summary of Recent Work, *Seism. Res. Lett.,* 68, 128-153.

Boore, D. M., W. B. Joyner, and T. E. Fumal (1994). Estimation of response spectra and peak accelerations from Western North America Earthquakes: An Interim Report, Part 2, U. S. Geological Survey Open-File Report 94-127, 40 pp.

Boore, D. M., W. B. Joyner, and T. E. Fumal (1997). Equations for estimating horizontal response spectral and peak acceleration from western North American earthquakes: A summary of recent work, *Seism. Res. Lett.,* **68**, 128-153.

Boore, D. M., and W. B. Joyner, (1991). Estimation of ground motion at deep-soil sites in eastern North America. *Bulletin of the Seismological Society of America,* **81** (6), 2167-2185.

Boore, D. M., and G. M. Atkinson (1987). Stochastic prediction of ground motion and spectral response parameters at hard-rock sites in eastern North America, *Bulletin of the Seismological Society of America,* **77**, pp. 440-467.

Borcherdt, R. D. (1994). Estimates of site-dependent response spectra for design (methodology and justification), Earthquake Spectra, 10, 617-654.

Brackman, T. (2005) ShakeMap Implementation for the Upper Mississippi Embayment, Thesis, University of Memphis, Department of Earth Sciences.

Campbell, K. W. (2002). Prediction of strong ground motion using the hybrid empirical method: example application to eastern North America, submitted to *Bulletin of the Seismological Society of America.*

Campbell, K.W. (1997). Empirical near-source attenuation relationships for horizontal and vertical components of peak ground acceleration, peak ground velocity, and pseudoabsolute acceleration response spectra, *Seism. Res. Lett.* **68**, 154-179.

Converse, A. M., and A. G. Brady (1992). BAP: Basic Strong-Motion Accelerogram Processing Software; Version 1.0" by, USGS Open-File Report 92-296A.

Dewey J. W., B. Glen Reagor, L. Dengler, and K. Moley (1995). Intensity distribution and isoseismal maps for the Northridge, California, earthquake of January 17, 1994, U. S. Geological Survey Open-File Report 95-92, 35 pp.

Dreger, D. S. and A. Kaverina, (2000). Seismic remote sensing for the source process and near-source strong shaking: a case study of the Hector Mine earthquake , *Geophys. Res. Lett.* 27, 1941-1944.

Eguchi, R.T., Goltz, J.D., Seligson, H.A., Flores, P.J., Blais, N.C., Heaton, T.H., and Bortugno, E. (1997). "The Early Post-Earthquake Damage Assessment Tool (EPEDAT)," *Earthquake Spectra*, Vol. 13, No. 4, Oakland, California, pp. 815-832.

EPRI (1991). Standardization of cumulative absolute velocity, EPRI TR100082 (Tier 1), Palo Alto, California, Electric Power Research Institute, prepared by Yankee Atomic Electric Company.

Electric Power Research Institute (1993). Guidelines for determining design basis ground motions. Palo Alto, Calif: Electric Power Research Institute, vol. 1 5, EPRI TR-102293.

Electric Power Research Institute (2004). CEUS Ground Motion Project. Palo Alto, Calif: Electric Power Research Institute, EPRI Final Report 1009684.

FEMA 222A (1994). NEHRP recommended provisions for the development of seismic regulations for new buildings, 1994 edition, Part 1 – provisions, Federal Emergency Management Agency, 290.

Field, E. H. P. A. Johnson, I. A. Beresnev, and Y. Zheng (1997). Nonlinear sediment amplification during the 1994 Northridge earthquake, Nature, 390, 599-602.

Field, E.H. (2000). A modified ground-motion attenuation relationship for southern California that accounts for detailed site classification and a basin-depth effect, *Bull. Seism. Soc. Am.,* **90**, S209-S221.

Frankel, A., Mueller, C., T. Barnhard, D. Perkins, E.V. Leyendecker, N. Dickman, S. Hansen, and M. Hopper (1996). National seismc-hazard maps: documentation, *U.S. Geol. Surv. Open-File Rept.* **96-352**.

Frankel, A. D., C. Mueller, T. Barnhard, D. Perkins, E. Leyendecker, N. Dickman, S. Hanson, and M. Hopper (1996). National seismic-hazard maps: documentation June 1996, *U.S. Geological Survey, Open-file Report* 96-532, 110.

Frankel, A. D., M. D., Petersen, C. S. Mueller, K. M. Haller, R. L. Wheeler, E. V. Leyendecker, R. L. Wesson, S. C. Harmsen, C. H. Cramer, D. M. Perkins, and K. S. Rukstales (2002). Documentation for the 2002 Update of the National Seismic Hazard Maps U.S. *U.S. Geological Survey, Open-File Report:* 02-420.
http://pubs.usgs.gov/of/2002/ofr-02-420/OFR-02-420.pdf

Hall, J. F., T. H. Heaton, M. W. Halling, and D. J. Wald (1995). Near-source ground-motions and its effects on flexible buildings, Earthquake Spectra, 11, 569-606.

Hartzell, S. H., S Harmsen, A. Frankel, D. Carver, E. Cranswick, M. Meremonte, and J. Michael (1998) First-generation site response maps for the Los Angeles region based on earthquake ground-motions, 88, 463-472.

Hauksson, E., P. Small, K. Hafner, R. Busby, R. Clayton, J. Goltz, T. Heaton, K. Hutton, H. Kanamori, J. Polet, D. Given, L. M. Jones, and D.J. Wald (2002). Southern California Seismic Network: Caltech/USGS Element of TriNet, *Seismol. Res. Let.,*

Hauksson, E., P. Small, K. Hafner, R. Busby, R. Clayton, J. Goltz, T. Heaton, K. Hutton, H. Kanamori, J. Polet, D. Given, L. Jones, and D. Wald (2001). Southern California Seismic Network: Caltech/USGS Element of TriNet, *Seism. Res. Lett.,* 72, no. 6.90-702.

Ji. C., D. V. Helmberger, and D. J. Wald (2004). A teleseismic study of the 2002 Denali, Alaska, earthquake and implications for rapid strong motion estimation, submitted to *Earthquake Spectra.*

Japan Meteorological Agency (1996). Note on the JMA seismic intensity, JMA report 1996, *Gyosei* (in Japanese).

Joyner, W. B. and Boore, D. M. (1988). Measurement, characterization, and prediction of strong ground-motions, in Proc. Conf. on Earthq. Eng. & Soil Dyn. II, Geotechnical vision, Am. Soc. Civil Eng., Park City, Utah, 43-102.

Joyner, W. B. and Boore, D. M. (1981). Peak horizontal accelerations and velocity from strong-motion records including records from the 1979 Imperial Valley, California, earthquake, 71, 2011-2038.

Kaka, S. I., and G. M. Atkinson (2004). Relationships between instrumental intensity and ground motion parameters in eastern North America. *Bulletin of the Seismological Society of America,* **94**, 1728 - 1736.

Kaka, S. I., and G. M. Atkinson (2005). Empirical ground-motion relations for ShakeMap applications in southeastern Canada & the northeastern United States, *Seismological Research Letters* (in press).

Kanamori, H. (1993). Locating earthquakes with amplitude: Application to real-time seismology, 83, 264-268.

Kanamori, H., and D. L. Anderson (1975). Theoretical basis of some empirical relations in seismology, *Bulletin of the Seismological Society of America*, **65**, 1073-1095.

Kanamori, H., P. Maechling, and E. Hauksson (1999). Continuous monitoring of ground-motion parameters, *Bull. Seism. Soc. Am.*, 89, 311-316.

Kanamori, H., E. Hauksson, and T. Heaton (1991). TERRAscope and CUBE project at Caltech, *EOS,* 72, 564.

Kanezashi, S., and F. Kaneko, (1997). Relations between JMA's measuring seismic intensity (MI) and physical parameters of earthquake ground-motion, OYO Technical Report, 1997, 85-96.

Kircher, C. A., R. K. Reitherman, R. V. Whitman, and C. Arnold, 1997. Estimation of earthquake losses to buildings, *Earthquake Spectra*, **13**, 703-720.

McGuire, R. K., and G. R. Toro (1987). Issues in strong ground motion estimation in eastern North America Proceedings from the Symposium on seismic hazards, ground motions, soil-liquefaction and engineering practice in eastern North America 361-374

Mori, J., H. Kanamori, J. Davis, E. Hauksson, R. Clayton, T. Heaton, L. Jones, and A. Shakal (1998). Major improvements in progress for southern California earthquake monitoring, 79, p. 217, 221.

National Institute of Building Sciences (NIBS), 1997. Earthquake Loss Estimation Methodology: HAZUS97 Technical Manual, Report prepared for the Federal Emergency Management Agency, Washington, D.C.

NIBS (1999), *HAZUS Technical Manual*, SR2 edition, Vols. I, II, and III, prepared by the National Institute of Building Sciences for the Federal Emergency Management Agency, Washington, D.C.

Newmark, N. M., and W. J. Hall (1982). Earthquake spectra and design, *Geotechnique*, 25, no. 2, 139-160.

Newmark, N. M., and W. J. Hall (1982). Earthquake Spectra and Design, Engineering Monographs on Earthquake Criteria, Structural Design, and Strong Motion Records, Vol. 3, Earthquake Engineering Research Institute, University of California, Berkeley, CA.

Pankow, K. L, and J. C. Pechmann (2003). Addedum to SEA99: A new PGV and revised PGA and pseudovelocity predictive relationship for extensional tectonic regimes, Submitted to *Bull. Seism. Soc. Am.*

Petersen, M. D. P., W. A. Bryant, C. H. Cramer, T. Cao, and M. Reichle, A. D. Frankel, J. J. Lienkaemper, P. A. McCrory, and D. P. Schwartz, (1996). Probabilistic Seismic Hazard Assessment For The State of California, *California Division of Mines and Geology Open-File Report* **96-08.**

Reasenberg, P., and D. Oppenheimer (1975). FPFIT, FPPLOT, and FPPAGE: Fortran programs for calculating and displaying earthquake fault plane solutions, *U. S. Geological Survey Open-File Report 75-739*, 109 pp.

Richter, C. F. (1958). Elementary Seismology. W. F. Freeman & Co.

Safak, E. (2000). A simple method to account for the effects of vertical loads on the horizontal seismic response of buildings, proceedings (CD-ROM), 6th International Conference on Seismic Zonation, Nov. 12-15, Palm Springs, California.

Scientists from the U.S. Geological Survey, Southern California Earthquake Center, and California Division of Mines and Geology (2000). Preliminary Report on the 10/16/1999 M7.1 Hector Mine, California Earthquake, *Seism. Res. Lett.*, 71, 11-23.

Scrivner, C. W., C. B. Worden, and D. J. Wald (2000), Use of TriNet ShakeMap to Manage Earthquake Risk, *Proceedings of the Sixth International Conference on Seismic Zonation*, Palm Springs.

Shakal, A., C. Peterson, A. Cramlet, and R. Darragh (1996). Near-real-time CSMIP strong motion monitoring and reporting for guiding event response, in *Proceedings of the 11th World Conference on Earth. Eng.*, Acapulco, Mexico.

Shakal, A., C. Peterson, and V. Grazier (1998). Near-real-time strong motion data recovery and automated processing for post-earthquake utilization, *Sixth Nat'l Conference on Earth. Eng.*, Seattle.

Shimuzu, Y. and Yamasaki, F., 1998, "Real-time City Gas Network Damage Estimation System–SIGNAL," *Proceedings of the 11th European Conference on Earthquake Engineering*, A.A. Balkema, Rotterdam.

Smith, W. H. F., and P. Wessel (1990). Gridding with continuous curvature splines in tension, *Geophysics* 55, 293-305.

Sokolov, V. Y. and Y. K. Chernov (1998). On the correlation of Seismic Intensity with Fourier Amplitude Spectra, *Earthquake Spectra*, Vol. 14, 679-694.

Somerville, P. G., N. S. Smith, R. W. Graves, and N. A. Abrahamson (1997). Modification of empirical strong ground-motion attenuation relations to include the amplitude and duration effects of rupture directivity, *Seism. Res. Lett.*, **68**, 199-222.

Somerville, P., N. Collins, N. Abrahamson, R. Graves, and C. Saikia (2001). Ground motion attenuation relations for the central and eastern United States, final report to U.S. Geological Survey.

Spudich, P., W.B. Joyner, A.G. Lindh, D.M. Boore, B.M. Margaris, and J.B. Fletcher, 1999, SEA99 - A revised ground-motion prediction relation for use in extensional tectonic regimes, *Bull. Seism. Soc. Am.*, 89, 1156-1170.

Street R., E. W. Woolery, J. Chiu (2004). Shear-wave velocities of the Post Paleozoic sediments across the Upper Mississippi Embayment, *Seismological Research Letters*, 75, 390-405.

Thio, H. K., and H. Kanamori (1995). Moment tensor inversion for local earthquakes using surface waves recorded at TERRAscope, *EOS*, Vol. 85, 1021-1038.

Toro, G. R., and R. K. McGuire (1987). An investigation into earthquake ground motion characteristics in eastern North America, *Bulletin of the Seismological Society of America*, 77, 468–489.

Toro, G. R., N. Abrahamson, and J. Schneider (1997). Model of strong ground motions from earthquakes in the central and eastern North America: best estimates and uncertainties, *Seismological Research Letters*, **68**, 41-57.

USGS (1999). An assessment of Seismic Monitoring in the United States: Requirements for an Advance National Seismic System, *U. S. Geological Survey Circular 1188*.

Wald, D. J., P. A. Naecker, C. Roblee, and L. Turner (2003). Development of a ShakeMap-based, earthquake response system within Caltrans, in *Advancing Mitigation Technologies and Disaster Response for Lifeline Systems*, J. Beavers, Ed., Technical Council on Lifeline Earthquake Engineering, Monograph No. 25, August 2003, ASCE.

Wald, D. J. (1999). Gathering of Earthquake Shaking and Damage Information in California, *Proceedings* of the 3rd US-JAPAN High Level Policy Forum, Yokohama, Japan.

Wald, D. and J. Goltz (2001). ShakeMap: A new Tool for Emergency Management and Public Information, *Proceedings* of the Los Angeles/Yokohama Disaster Prevention Workshop, Yokohama, Japan, November, 2001.

Wald, D., L. Wald, J. Dewey, V. Quitoriano, and E. Adams (2001). Did You Feel It? Community-Made Earthquake Shaking Maps, *U.S. Geological Survey Fact Sheet* 030-01.

Wald, D., L. Wald, J. Goltz, B. Worden, and C. Scrivner (2000). "ShakeMaps" — Instant Maps of Earthquake Shaking, U.S. Geological Survey Fact Sheet 103-00.

Wald, D, L. Wald, B. Worden, and J. Goltz (2003). ShakeMap — A Tool for Earthquake Response, U.S. Geological Survey Fact Sheet 087-03.

Wald, D. J., and T. H. Heaton, and K. W. Hudnut (1996). Rupture history of the 1994 Northridge, California earthquake from strong-motion, GPS, and leveling data, *Bull. Seism. Soc. Am.*, 86, S49-S70

Wald, D. J., T. Heaton, H. Kanamori, P. Maechling, and V. Quitoriano (1997). Research and Development of TriNet "Shake" Maps, *EOS*, 78, No. 46, p F45.

Wald, D. J., V. Quitoriano, T. H. Heaton, H. Kanamori (1999b). Relationship between Peak Ground Acceleration, Peak Ground Velocity, and Modified Mercalli Intensity for Earthquakes in California, *Earthquake Spectra*, Vol. 15, No. 3, 557-564.

Wald, D. J., V. Quitoriano, T. H. Heaton, H. Kanamori, C. W. Scrivner, and C. B. Worden (1999a). TriNet "ShakeMaps": Rapid Generation of Peak Ground-motion and Intensity Maps for Earthquakes in Southern California, *Earthquake Spectra*, Vol. 15, No. 3, 537-556.

Wald, D. J., V. Quitoriano, L. Dengler, and J. W. Dewey (1999c). Utilization of the Internet for Rapid Community Intensity Maps, *Seism. Res. Letters*, 70, No.6, 680-697.

Wald, D. J. (1999). Gathering of Earthquake Shaking and Damage Information in California, *Proceedings of the 3rd US-JAPAN High Level Policy Forum*, Yokohama, Japan.

Wald, D., L. Wald, J. Goltz, B. Worden, and C. Scrivner (2000). "ShakeMaps" — Instant Maps of Earthquake Shaking, *U.S. Geological Survey Fact Sheet* 103-00.

Wessel, P. and W. H. F. Smith, (1991). Generic Mapping Tools, *EOS*, Vol. 72, 441.

Wills, C. J., M. D. Petersen, W. A. Bryant, M. S. Reichle, G. J. Saucedo, S. S. Tan, G. C. Taylor, and J. A. Treiman (2000). A site-conditions map for California based on geology and shear wave velocity, *Bull. Seism. Soc. Am.* ,**90** , S187-S208.

Wood, H. O. and Neumann (1931). Modified Mercalli intensity scale of 1931, Bull. Seism. Soc. Am. 21, 277-283. Yamakawa, K. (1998). The Prime Minister and the earthquake: Emergency Management Leadership of Prime Minister Marayama on the occasion of the Great Hanshin-Awaji earthquake disaster, *Kansai Univ. Rev. Law and Politics*, No. 19, 13-55.

Wu, Y, M. W. H. K. Lee, C. C. Chen, T. C. Shin, T. L. Teng, and Y. B. Tsai (2000). Performance of the Taiwain Rapid Earthquake Information Release System (RTD) during the 1999 Chi-Chi (Taiwan) earthquake, *Seism. Res. Lett.*, **71**, 338-343.

Wu, Y. M., T. C. Chin, and C. H. Chang (2001). Near real-time mapping of peak ground acceleration and peak ground velocity following a strong earthquake, *Bull. Seism. Soc. Am.* ,**91**, 1218-1228.

Wu, Y. M., T. L. Teng, T. C. Shin, and N. C. Hsiao (2003). Relationship between peak ground acceleration, peak ground velocity and Intensity in Taiwan, *Bull. Seism. Soc. Am.* ,**93**, 386-396.

Youngs, R. R., S.-J. Chiou, W. J. Silva, and J. R. Humphrey (1997). Strong ground-motion relationships for subduction zones, *Seism. Res. Letters*, 68, No.1, 58-73.

APPENDIX A. Regression Relationships

The following ground-motion attenuation or regressions are available in the ShakeMap package. They may be selected as the de facto regression for a region, used automatically used for events within a certain magnitude and depth ranged, or manually selected for specific events or scenario events.

Boore and others (1997), PGV modified by Newmark & Hall (1982)	So. California, default regression
Boatwright and others (2003)	No. California, default regression
Atkinson and Boore (2002)	Scenarios only (Cascadia region)
Somerville (1997)	Scenarios only (directivity effects)
Youngs and others (1997)	Washington and Alaska (depth at least 41 km)
ShakeMap Small Regression	All regions (M<5.3)

The regressions calculate both random and peak component values of the estimated parameters. The equations given are for the mean values. We derive the peak values by scaling up the mean value by 15 percent (Joyner, Campbell, personal communication.) Note that the site correction components of the regressions are ignored unless specified; for those without site corrections, the Borcherdt (1994) site correction method is used.

Boore and others 1997 (BJF97)

This attenuation model is used as the default relation in southern California for all events with magnitude ≥ 5.3. The relation has the form:

$$\ln (Y) = B1 + B2(M\text{-}6) + B3(M\text{-}7)2 - B5 \ln R \qquad (A.1)$$

where

> Y is either PGA or PSA in g
> M is the magnitude
> R = sqrt(Rjb2 + h2), see below

Rjb is the "Joyner-Boore" distance to the surface projection of the fault, in km. This model assumes a shallow fault and uses only a 2D fault model with no depth term.

Values for B1-B5 and h are given below. BJF97 does not predict 3 s. PSA; we use the coefficients for 2 s. PSA. The factors for average slip type are used for triggered events. However, the slip type may be specified for scenario earthquakes in the event file, in which case the regression will apply the appropriate coefficients.

Slip type	PSA Period (s)	B1	B2	B3	B5	h (km)
Strike-slip	PGA	-0.313	0.527	0.000	-0.778	5.57
	0.3	0.598	0.769	-0.161	-0.893	5.94
	1.0	-1.133	1.036	-0.032	-0.798	2.90
	3.0	-1.699	1.085	-0.085	-0.812	5.85
Reverse	PGA	-0.117	0.527	0.000	-0.778	5.57
	0.3	0.803	0.769	-0.161	-0.893	5.94
	1.0	-1.009	1.036	-0.032	-0.798	2.90
	3.0	-1.801	1.085	-0.085	-0.812	5.85
Average	PGA	-0.242	0.527	0.000	-0778	5.57
	0.3	0.700	0.769	-0.161	-0.893	5.94
	1.0	-1.080	1.036	-0.032	-0.798	2.90
	3.0	-1.743	1.085	-0.085	-0.812	5.85

PGV is derived from PSA (1.00) using the Newmark and Hall 1982 relation (NH82). See Section 2.1.1.2. For comparison purposes, we also provide an earlier PGV regression relation using Boore and others (1982):

$$\log PGV = a + b(M-6) - d \log R + k R \qquad (A.2)$$

a	2.09
b	0.49
d	-1.00
k	-0.0026
e	-0.45
h	4.00 km

Boatwright and others 2003 (Boatwright03)

This attenuation model is used as the default relation in northern California for all events with magnitude $\geq$ 5.3. The relation has the form:

[TBS]
(A.3)

Newmark and Hall 1982 PGV Relation (NH82)

In order to conform with previous HAZUS studies, we derive peak ground velocity (PGV) from the 1.0 s spectral acceleration with the relationship of Newmark and Hall (1982).

$$PGV = PSA (1 s) * 37.27 * 2.54 \qquad (A.3)$$

where PSA is in g and PGV is in cm/s.

Few regressions have up-to-date PGV coefficients available. Hence, this relation is used in all online events and scenarios except for the ShakeMap Small Regression, which has its own PGV relation (See 2.1.1.x). For testing purposes, the PGV regression of Boore and others (1982) is available for scenarios along with the BJF97 model (See 2.1.1.1.)

Pankow and Pechman 2002
[TBS]
(A.4)

Atkinson and Boore 2003 (AB03)

This attenuation model is available for use in scenarios in the Cascades region or other deep-event subduction regions. Event depth is required for this regression, as well as event type (interface or intraslab). Because this regression normally used for deep earthquakes, only hypocentral distance is used; finite faults are not supported. This relation also uses a custom site correction (see below).

The relation has the form:

$$\log 10\,(Y) = c1 + c2\,M + c3\,h + c4\,R - g\,\log 10\,R \qquad\qquad (A.5)$$

Y is PGA or PSA in cm/s^2
M is the magnitude
R = sqrt (Rhypo2 + (0.00724 * 10(0.507 M))2)
g = 10(1.2 – 0.18 M) for interface events
  = 10(0.301 – 0.01 M) for intraslab events

Magnitude is capped at 8.5 for interface events, or 8.0 for intraslab events. Rhypo is the hypocentral distance. Values for c1-c5 are given below. PGV is derived from PSA (1.00) using the NH82 relation.

Event type	PSA Period (s)	C1	C2	C3	C4	C5
Interface	PGA	0.0	2.991	0.0352	0.00759	-0.00206
	0.3	2.5	2.525	0.148	0.00728	-0.00235
	1.0	1.0	2.144	0.134	0.00521	-0.00110
	3.0	0.33	2.301	0.0224	0.00012	0.0

	PGA	0.0	-0.0471	0.691	0.011	-0.00202
Intraslab	0.3	2.5	0.00544	0.7727	0.00173	-0.00178
	1.0	1.0	-1.0213	0.8789	0.00130	-0.00173
	3.0	0.333	-3.7001	1.1169	0.00615	-0.00045

The Atkinson and Boore (2003) regression uses a custom nonlinear site correction that replaces the default correction.

This site correction is of the form

log10 Y(soil) = log10 Yrock + sl (C5 Sc + C6 Sd + C7 Se)   (A.6)

Sc, Sd, and Se determine the soil velocity (Vs30) bin for the site:

    Sc = 1, Sd = Se = 0 if Vs > 360 m/s
    Sd = 1, Sc = Se = 0 if 180 m/s <= Vs < 360 m/s
    Se = 1, Sc = Sd = 0 if Vs < 180 m/s

and sl is a nonlinearity factor:

sl = 1 – (f-1) (PGArx – 100) / 400
   =1 if PGArx < 100 or f < 1
   = 0 if PGArx > 500

f is the frequency in Hertz (0 for PGA), PGArx is the predicted 'rock value' PGA in %g [check this] at the site. The values for C5-C7 are independent of event type and are given below.

Period (s)	C5	C6	C7
PGA	0.19	0.24	0.29
0.3	0.13	0.37	0.38
1.0	0.10	0.30	0.55
3.0	0.10	0.25	0.36

Somerville and others 1997 (Somerville97)

This attenuation model is identical the Boore and others (1997) model modified by the Somerville and others (1997). PGV is derived from PSA (1.00) using the NH82 relation. This model has recommended modifications that can be applied to existing attenuation relationships to explicitly add directivity in a deterministic sense to large strike slip events (magnitude range 6.0 – 6.5). A fault file is required, and it is assumed that the fault is a simple vertical strike slip single-segment fault defined by the endpoints.

The directivity correction at a site is of the form:

Ydirec = Y e(d)
d = (C1 + C2 s/L cos theta) Tr Tm   (A.7)

where

   Y is the original ground-motion parameter (in g)
   s/L is the length ratio (fraction of fault along strike that ruptures toward the
site)
   L is the fault length
   theta is the azimuth angle between the fault plane and the raypath to the site
   C1 and C2 are given below:

Parameter	Period in Somerville model (s)	C1	C2
PGA or PSA (0.3 s)	0.5	0	0
PGV or PSA (1.0 s)	1.0	-0.192	0.423
PSA (3.0 s)	3.0	-0.605	1.333

Note that the parameters in Somerville and others (1997) do not correspond completely to the ShakeMap parameters. The closest or most equivalent parameters have been used.

The directivity parameter d is further modified by a linear taper dependent on distance and magnitude given in Abramson (2000):

   Tr = 1 – (R-30) / 30 if 30 km <= R < 60 km   (A.8)
         = 1 if R < 30
         = 0 if R > 60

   Tm = 1 + (M – 6.5)2 if 6.0 <= M < 6.5   (A.9)
         = 0 if M < 6.0
         = 1 if M > 6.5

To date, we have not included this correction in the online ShakeMap system. Directivity is typically included implicitly in most regressions, that is, they contain data that represent the average directivity as recorded over a wide range of faulting directivity situations. Hence, by employing such a regression directivity is included in the empirical ground-motion estimates in an average sense.

In practice there are limitations to the explicit directivity approach of Somerville97. First, the assumption of a single linear fault segment is typically violated by large earthquakes, including the 1992 Landers, California (M7.3) and 2002 Denali, Alaska (M7.9) events, where total fault curvature, or change in strike reached 25-30 degrees. These relations require the angle with respect to the rupture direction, and the latter changes significantly *during* the rupture. Secondly, it has not yet been ascertained (mostly due to limited data) whether these recommended directivity functions adequately represent directivity from such large events. For example, using these functions, both ends of a 200 km bilateral rupture experience no directivity, yet intuitively, both points experience directivity due to a 100 km fetch of rupture coming toward each station. Finally, for rapidly determined ShakeMaps, directivity cannot be applied without a reasonable constraint on the rupture location and dimensions, which is not available in near-real time.

It is hoped that directivity for a large earthquake will be sample observational and hence will be locally constrained upon interpolation. Further improvement to the empirically-based predictive aspects of ShakeMap might include a azimuthally-dependent term to the bias correction, capable of adding directivity in real-time based on direct event-specific observations.

Youngs and others 1997 (Youngs97)

This attenuation model is used for the Washington and Alaska ShakeMap regions and for other subduction zones. Event depth is required for this regression, as well as event type (interface or intraslab). Because this regression normally used for deep earthquakes, either hypocentral distance of distance to a 3D fault model can be used. This model is specified by sets of planar segments (quadrilaterals), each planar segment joined at a common side. Each quadrilateral segment is defined in the fault file by four (coplanar, noncollinear) corner points. One or two planar segments should be sufficient for most cases.

The relation has the form:

$$\log(Y) = 0.2418 + 1.414\,M + C1 + C2\,(10 - M)3 \\ + C3\log(Rrup + 1.7818\,e(0.554\,M)) + 0.00607\,H \\ + 0.3846\,Zt \qquad (A.10)$$

Y is PGA or PSA\ in g
M is the magnitude
Rrup is the hypocentral distance or distance to fault, described above
H is the hypocentral depth
Zt = 1 for intraslab events, 0 otherwise

Values for c1-c5 are given below. PGV is derived from PSA (1.00) using the NH82 relation.

Parameter	C1	C2	C3
PGA	0	0	-2.552
PSA (0.3 s)	0.246	-0.0036	-2.454

| PSA (1.0 s) | -1.736 | 0.0064 | -2.234 |
| PSA (3.0 s) | -4.511 | -0.0089 | -2.003 |

ShakeMap Small Regression (Small)

The ShakeMap Small Regression is a modified form of the attenuation relationship for small events described in Wald and others (1999a) extending the event database to 2002. It is used as the default regression for events with magnitude below 5.3. The relation has the form:

$$\log 10\ (Y) = B1 + B2(M-6) - B5\ \log 10\ R \qquad\qquad (A.11)$$
where

> Y is PGA or PSA in cm/s^2 or PGV in cm/s
> M is the magnitude
> R = sqrt(Rjb2 + h2), see below
> h = 6.00 km

Rjb is the "Joyner-Boore" distance to the surface projection of the fault, in km. This model assumes a shallow fault and uses only a 2D fault model with no depth term. Values for B1-B5 are given below.

Parameter	B1	B2	B5	Sigma
PGA	4.037	0.572	-1.757	0.836
PGV	2.223	0.740	-1.386	0.753
PSA (0.3 s)	3.354	0.746	-1.827	0.842
PSA (1.0 s)	2.197	0.959	-1.211	0.988
PSA (3.0 s)	0.980	0.909	-0.848	1.082

Note that standard deviation values (sigmas) are total sigma defined in log10-amplitude space.

Depth to Basement

We have coded the depth of basement correction recommended by Field (2002). This model was developed using the Boore and others (1997) attenuation model but may be used for any relation. It is meant for use in scenarios only. The correction is applied to each grid point after interpolation to a fine grid, analogous to the site correction step.

By specifying a map of the depth to basement, the resulting ground-motion is modified by an amplification factor

$$Y basin = Y\ e(A\ d + B) \qquad\qquad (A.12)$$

where Y is the non-basin ground-motion (for PGA, PGV, or PSA), d is the basin depth in km, and A and B are parametric constants:

Parameter	A	B
PGA	6.7 x 10-5	-0.14
PGV	12.0 x 10-5	-0.25
PSA (0.3 s)	5.7 x 10-5	-0.12
PSA (1.0 s)	12.0 x 10-5	-0.25
PSA (3.0 s)	11.0 x 10-5	-0.18

Currently, this is functional in the Los Angeles basin region using the SCEC Southern California basin model (Magistrale and others, 2000), but we do not use it for the online generation of ShakeMaps. In part, this is because this correction is not that well established, nor are the basin depths well constrained, but more important, we have sufficient station sampling in the urban basin regions of to adequately represent deep basin effects observationally. That is, any data above a basin records all basin effects at that point. Interpolated values at adjacent points within the basin using that data naturally also reflect such effects. Hence, having representative sites in basins, near basin margins, and on rock will provide a firm basis for our interpolation, which is only otherwise constrained by shallow site amplification terms based on 30-m shear velocity estimates. Lacking representative observed values would naturally lead to poor representation of any potential 3-D amplification effects given the 1-D site corrections we apply; the greater the spatial separation, the greater the inference.

However, the basement depth correction term is useful for comparisons of ground-motion effects for scenario earthquakes in the region. This option can be easily configured prior to running a Scenario so we retain it for such exercises.

Toro *et. al.* 1997

Toro *et. al.* (1997) developed an attenuation relationship for Eastern North America based on the stochastic ground motion model. Two separate attenuation models were developed: 1) the Mid-Continent region which includes areas north of the Tennessee/Mississippi border and the northern half of Arkansas and 2) the Gulf Coastal Plain region representing the southern half of Arkansas and areas south of Tennessee (Toro *et. al.*, 1997). The model for the Mid-Continent region is used in creating ShakeMaps and the equation (A.13) is shown below.

The attenuation equation as given by Toro *et. al.*, (1997) is:

$$\ln(Y)=C_1+C_2(M-6)+C_3(M-6)^2-C_4\ln R_M-(C_5-C_4)\max[\ln(R_M/100),0]-C_6 R_M \qquad (A.13)$$

where,

   ln Y is peak ground acceleration or spectral acceleration in units of g,
   $R_M = \sqrt{R_{jb}^2 + C_7^2}$

   R_{jb} = distance to surface expression of fault plane (as defined in Abrahamson and Shedlock, 1997)

and

M is moment magnitude.

Coefficients for determining peak ground acceleration and pseudo-acceleration are shown below.

Coefficients for Mid-continent and Moment Magnitude (M) (Toro, 1997)							
Freq. (Hz)	C1	C2	C3	C4	C5	C6	C7
0.5	-0.74	1.86	-0.31	0.92	0.46	0.0017	6.9
1.0	0.09	1.42	-0.20	0.90	0.49	0.0023	6.8
5.0	1.73	0.84	0.00	0.98	0.66	0.0042	7.5
PGA	2.20	0.81	0.00	1.27	1.16	0.0021	9.3

The attenuation relationship for Toro *et. al.,* (1997) was configured to return peak ground motion values on hard rock with a reference velocity of approximately 1800 m/s. Distance is defined as R_{jb} (as defined in Abrahamson and Shedlock, 1997). The ShakeMap routines scale the values to return %g and scale up the values by 15% to estimate a maximum value rather than a random component (Wald et. al., 2004). Values were calculated for peak ground acceleration, pseudo-acceleration (PSA 5% damped) 2.0, 1.0, and 0.30 seconds (Toro *et. al.,* 1997). Peak ground velocity coefficients are not available (Toro, personal communication), and velocity was computed from 1-Hz PSA, in keeping with HAZUS studies (Wald *et. al.,* 2004), using the Newmark-Hall (1982) equation:

$$PGV = (PSA)(37.27)(2.54)$$

where,

PSA is pseudo-acceleration at 1 s. in g,

and

PGV is in cm/s.

Atkinson and Boore 1995

Atkinson and Boore (1995) used the semi-empirical stochastic approach, using a two-corner frequency source model to estimate hard rock ground motions. The polynomial equation of the modeled data over predicted for magnitudes below six and the use of published table values was highly recommended (Kaka, personal communication).

The attenuation relationship module for Atkinson and Boore (1995) was created by the ShakeMap working group (Quitoriano, personal communication). The polynomial expression was replaced by smoothed table values (Wald, personal communication) of peak ground acceleration, peak ground velocity and pseudo-acceleration (5% damped) at 2.0, 1.0, and 0.30 seconds for a given magnitude and distance. The resulting values were multiplied by 0.15 to get a maximum rather than random component (Wald *et. al.,* 2004). This regression used hypocentral distance (R_{hypo}). Magnitude was constrained between 2.5 - 7.5 and R_{hypo} between 10 km and 1000 km. The regression assumes base rock is NEHRP soil type C or 760 m/s and has a custom site correction method (site_correct_ab02) (Wald *et. al.,* 2004):

10**(c5*sl*Sc + c6*sl*Sd + c7*sl*Se)

where,

  sl is a nonlinearity factor

and

  Sc, Sd, and Se are NEHRP soil shear wave velocities.

Kaka and Atkinson (2005)

  Kaka and Atkinson (2005), used empirical and modeled data to developed an attenuation relationship for pseudo-acceleration (5% damped) at frequencies of one, two, five, and ten hertz, peak ground acceleration in cm/s^2 and peak ground velocity in cm/s, for the central and eastern United States. Peak ground motion equations were obtained by a simple linear regression of the assembled data (Kaka and Atkinson, 2005). The general form for the peak ground motion equation is:

$$\text{Log } Y = C_1 + C_2 (M-4) + C_3 (M-4)^2 + C_4 \text{ Log } R + C_5 R \quad 3.1 \tag{A.14}$$

where

  Y is the vertical component ground motion parameter (PGV in mm/s and PGA/PSA(f) in cm/s^2),

  R is hypocentral distance (R_{hypo}) in km,

and

  M is a moment magnitude.

Coefficients for determining peak ground acceleration, peak ground velocity and pseudo-acceleration are shown below.

Coefficients for Quadratic Equation (Kaka and Atkinson, 2005)					
Freq. (Hz)	C_1	C_2	C_3	C_4	C_5
1.0	0.209	1.047	0.015	-0.854	-7.091e-6
2.0	1.185	1.068	-0.060	-0.963	-1.845e-4
5.0	1.891	0.943	-0.074	-0.922	-9.77e-4
10.0	2.524	0.825	-0.061	-1.094	-0.0013
PGA	2.779	0.855	-0.050	-1.433	-7.563e-4
PGV	1.496	0.899	0.029	-1.268	-9.146e-5

  The attenuation relationship for Kaka and Atkinson, (2005) is configured to return a random vertical component on rock, with a reference velocity of approximately 1800 m/s. Distance was defined as R_{hypo} but the module is presently configured to use R_{JB}. Peak ground velocity is converted from mm/s to cm/s. The ShakeMap routine returned %g and scaled up the values by 15% to estimate a maximum value rather than a random component (Wald et. al., 2004). Values are calculated for peak ground velocity and pseudo-acceleration (5% damped) at 1.0, 0.10 and 0.20 seconds.

  Kaka and Atkinson's (2005) equation for attenuation returns a random vertical component. A conversion from calculating vertical peak ground velocity (PGV_V) to horizontal peak ground velocity (PGV_H) was needed. For central and eastern United States the average horizontal to vertical ratio for hard rock is Kaka and Atkinson (2005):

$$\frac{PGV_H}{PGV_V} = 1.21 \qquad\qquad\qquad\qquad\qquad\qquad\text{(A.15)}$$

substituting into the above equation (A.14) and solving for $LogPGV_H$ gives:

$$LogPGV_H = C_1 + C_2(M-4) + C_3(M-4)^2 + LogR + C_5 R(3.1) + Log(1.21). \qquad\text{(A.16)}$$

Therefore, to determine the horizontal component on rock for calculations using the Kaka and Atkinson, (2005) attenuation relationship, the above equation (A.16) was used.

APPENDIX B. Supplemental Documents

ShakeMap Fact Sheet

http://pubs.usgs.gov/fs/fs-087-03/

ShakeCast Information Sheet

http://www.shakecast.org/pdf/ShakeCastIntroduction.pdf

Introduction to ShakeCast

http://www.shakecast.org/pdf/ShakeCastIntroduction.pdf

Using ShakeMap in HAZUS

http://earthquake.usgs.gov/shakemap/sc/shake/ShakeMap2HAZUS.html

INDEX

www.ingramcontent.com/pod-product-compliance
Lightning Source LLC
Chambersburg PA
CBHW080251180526
45167CB00006B/2493